CARING
as a
CARER

GEORGE MAILATH

CONTENTS

DEDICATION

*This book is dedicated to the memory of Jenny Mailath who,
after seven years of having suffered the effects of Multiple
System Atrophy has, on the 15th September 2014 joined the
ranks of the increasing number of those MSA Angels who faced
the inevitable fate of those suffering this relentlessly cruel terminal
and untreatable disease.*

*May God bless the persistence on the part of the many engaged in
the seemingly never ending efforts conducting research into the
answer of the cause of this disease and ultimately discover a cure.*

*May God also bless those carers who are dedicated to giving comfort
to all those suffering this and many other similarly cruel conditions and
may our Jenny and her fellow angels by the grace of God be granted
eternal rest.*

INTRODUCTION

*I*t was a terrible night.

Heavy rain and the sort of wind which make you concerned that the windows of the house could cave in at any time.

It was a night when sleeping was out of the question and so it was a time for thinking.

Over the last five and a half years my thinking involved little else than my wife Jenny and her having been diagnosed with Multiple System Atrophy (MSA), the relentless incurable condition with which I have not and will be unlikely to ever come to terms.

The contents of this entire document is primarily relating my experiences, the result of the most substantial learning curve I have ever encountered, and is intended for those who similarly to myself became the primary carer or assisted in caring for someone they love.

During the course of recording these experiences I have come to the conclusion that if I could have learned as much of the disease itself as I did of many other related factors not the least of which are a more intimate knowledge of myself and the inevitable conclusion that when, one of a couple, be it a marriage or a partnership, is diagnosed with MSA or most other similar terminal diseases the effect is devastating for both.

When I mention carers in this book I address my comments to those myriads of dedicated people who are not professional carers, in other words not carers who are employed and remunerated for their work.

That is not to say that professional carers cannot benefit from the contents of this book as working in any professional environment no matter how skilled and how knowledgeable we are, we continue to run across situations from which we can and do learn.

Naturally where I mention such things as training, carers employed by various institutions and facilities undergo training as part of the responsibility of the employer. As a result they are likely to be of benefit to volunteer carers by their training and their experience as part of their profession and are of considerable help to volunteers working alongside them.

I considered it necessary to increase the number of chapters and events to a number greater than originally intended.

This was recognising that the period from the delivery of the diagnosis up and until the loss of the battle by Jenny, did contain episodes which were very specific as being different and containing some relevant events.

Though unpleasant and sad experiences, they each in their own right resulted in several changes of behaviour as well as new and important practices which heavily influenced the caring procedures.

Carer's roles will relate to such things as the location, the circumstances, the hours involved as well as the activities to be dedicated to full-time or regarded as a casual activity.

It is my earnest hope that those reading this book will not only be those carers who are currently involved in providing care, either individually to someone they love, or out of generosity to a friend or neighbour, but in the main those currently on the threshold of starting to provide voluntary care contemplating caring as *a career.*

This book will be full of examples where I have encountered situations which held no answers for me and when I commenced to look after my wife Jenny at the outset of her illness I found more questions than answers.

So in many respects I was somewhat like a pioneer remembering of course that Jenny's illness was at that time hardly heard of and there would still be a lot of people today who would not know what multiple system atrophy really is.

As you read this book many of the areas which represented a challenge because of the absence of any knowledge in my case, will now at least have many of those already answered making your task of caring a little easier.

Before getting down to caring roles, I have thought long and hard during the early days of starting this book, whether and to what extent can we suggest that a substantial part of what is contained herein, whilst clearly concentrating on MSA, will, I believe apply to almost any and all caring provided to sufferers of terminal conditions as well as other serious conditions which may well be curable.

To be able to turn that comment into reality it will clearly be important for the reader as an active carer, to discuss any of the content about which there may be some doubt with the general practitioner of the patient who will no doubt be in a position to advise whether or not the particular comment made applies in that individual case.

Over the years I found by talking to a number of people involved in caring that, like any other such activity, relative knowledge may have been obtained through training, through actual experience and in some cases purely from a desire to help someone seriously ill, during which some experience may have been gained.

Partly because there have been periods of time which included before and after events as they occurred as well as to ensure that the contents are reasonably easy to follow, the work is dissected into a number of chapters or events recognising the importance of the effect both psychologically and physically on the patient, the carers as well as the immediate people associated with those mentioned in the text but not necessarily taking an active role in caring.

I point out at various times throughout the book that apart from some episodes where I have come into contact with people in a similar situation to Jenny but where I was not involved in the care, much of what I have witnessed would have had an inevitable and beneficial influence on some of my subsequent actions.

I regard some training as imperative and, that it can come from formal training, or your association with trained carers.

An example would typically be a situation where your patient is receiving care from outside sources normally provided by approved organisations most of which come from the various churches and other similar organisations.

These practitioners will, by their very actions and their conversation provide you with knowledge and information which will likely equip you with some training on which you will be able to build from other sources.

I found it both necessary and beneficial to start with a time period approaching the beginning of Jenny's illness as in retrospect, many things were happening and can well, after the event, be considered to have had an effect to some extent on most if not all of the people who are featured in this book.

It is important now to explain the manner in which I decided to write this book and what made me give that matter a great deal of thought.

It occurred to me from the outset that if I purely talked about my experiences as a carer with the accent being on such things as symptoms, the attempts at certain therapies and their effect on Jenny, there would be a tendency for the book to turn into some sort of textbook. I do not believe this would be appropriate and that there would be a likelihood of the majority of people reading the book finding it uninteresting and lifeless.

I decided instead, to tell it as it was, like a story which would in no way detract from what value may be found in references to the activity of caring, including all of the joys and sadnesses involved in providing care to someone you love and it would certainly not hurt if some of the emotion involved in the reality of caring would touch at least some of those reading it.

Clearly, the intent behind this book is to provide a volume of information, principally meant for other people who have found themselves in a similar situation as I have. It should also be clear that whatever is contained in this book should at best be regarded as examples, ideas and some experiments most of which are personal experiences encountered by me in my day-to-day activities as a carer.

As I will keep stressing, it is not backed by any professional qualifications, except where I may use quotations from other works.

It is essential to state that, as this book was largely written after Jenny's passing, from notes taken "on the run" during Jenny's illness and as most of these notes were undated, any appearance of dates tend to be rare.

Given that a great many details contained in the book refer to specific episodes which are described as if occurring at the time however when discussing choking for example we may talk about some episodes of choking which happened at different times.

I therefore considered that, except on certain occasions dates became somewhat irrelevant.

It is essential to understand that this is an account of periods of our life some aspects of which may not even bear any similarity to someone else's reality. I do hope however that the many different events and experiences contained relating purely to the activity of caring will strike an accord with most reading this work.

What now follows is the main body of the book which will discuss in great detail Jenny's journey with MSA, my journey with her and the journey of Jenny's sister Jeanette who assisted with her care, also Jenny Chin the wife of Barry, Jenny's older brother who also assisted. Jenny's brother Ray from Brisbane who gave up a lot of his time to assist with Jenny's care and whose calm and rational comments were always very welcome, as well as other family members. Maxine our neighbour to whom I continually referred to as our guardian angel and who, in spite of being a very busy person, spent an extraordinary amount of time sitting at Jenny's side. Dr David Barker who was not only our Dr but a very good friend without whose counsel our time would have been a great deal more difficult. Fr David our parish priest who was as much a good friend as a great parish priest. *May God bless them all!*

I am assuming that as this book circulates among the MSA community it will be read by not only the carers but I would hope members of the families of all MSA sufferers because the many misconceptions that have at times made the role of carers even more difficult will probably be dispelled. What is contained in this book was my actual experience as a carer. I cover such things as the quality of care given by nursing homes, the dedication on the part of the staff and certainly not forgetting the many people working for and with organisations such as the Anglican Retirement Villages people who looked after Jenny throughout the time before her entry into the nursing home.

"*Before the Beginning*"

S o, the first period covered in this chapter presents Jenny and I as an exceptionally happy couple, reasonably comfortable in terms of worldly goods and as Jenny was unable to bear a child and both of us being animal lovers over this period of our lives, we owned a dog for the better part of the time and for the remaining periods we were owned by cats, one called Cuddles was with us for 16 years and died of cancer and Chloe, who was with us for ten years while Jenny lived and is still with me now, showing very few signs of ageing but, having lived with us through the remainder of time to the present, her demeanour, her character and personality have undergone distinct changes.

Our life, living in Blackbutt which is in the Shellharbour area of New South Wales, commenced soon after our arrival from Canberra in December 1992, following the death of Jenny's mother as at that time Jenny had three members of her family residing in the Wollongong / Illawarra area, Jenny felt that the move would be a right one bringing us closer to members of her family.

As part of our move, we sold our accounting practice in Canberra and soon after arriving at what was to be our new home, I started a new practice which, having been over the age of 60 was by some considered to be somewhat ambitious.

As Jenny was accustomed to working, we decided that she should work as part of the new practice. She initially performed the duties of receptionist and as the practice grew somewhat, took over the mantle of practice manager which, because of her sunny disposition, outstanding people skills and her ability to appear to take a personal interest in everybody she met; this change certainly was for the better especially so far as the practice was concerned.

This simpler life contrasted with my situation during our years in Canberra where, as well as having the practice I was also very active in politics and public affairs. I was preselected and stood for Federal election in 1975 and 1977 and the fact that neither of those two attempts were successful, in hindsight I am quite pleased about as I am certain that I would have had no joy being a member of Federal Parliament. I also became very active in the affairs of small business both as a professional and in a political sense.

I was instrumental in establishing the branch of the Australian Small Business Association (ASBA) in Canberra and subsequently took over as Federal President of the Association and was also elected Chairman of the Australian Free Enterprise Foundation both occurring in December 1988. So, when establishing the new practice after our arrival in the Wollongong area, I was firmly resolved not to allow the practice to grow too large and also to refrain from being involved in other important but time-consuming activities.

I have never heard Jenny complain about the lifestyle we lived in Canberra even though quite obviously my frequent absences when needing to attend meetings interstate as well as activities involving a great deal of time locally must have made her existence rather difficult.

Our new life after the practice settled in was quite ideal because although we worked very hard we had resolved to leave ample time for "smelling the flowers" and as frequently as possible we would create long weekends and would travel to Sydney or less frequently Melbourne, stay at nice motels and attend shows if there was something which took our fancy.

Attending a wedding with no fear of the future.

We were both enjoying good health and of course I must not overlook my passion for golf, the quality of which in my case was never equal to the level of my passion.

So, looking back, these were exceptionally memorable times and there was never any doubt in my mind that our marriage was certainly good enough to be used as an example of a perfect marriage.

I reluctantly use the hackneyed phrase; "all good things come to an end" but in many respects the basic meaning of the phrase does have some applications here. It is rather sad to reflect that this was about the time when Jenny and I had commenced some fairly serious discussions on our plans for retirement.

Because of my early activities after arriving in Australia and after having completed what was then a contract with the government which compelled all new arrivals under the displaced persons scheme to give two years of their time and work where the government directed, I learned a little bit about Australia, from the point of view of residents of the Bush (the area where I spent those two years of contract), and I was keen to visit as many places in Australia as I was able. Circumstances at the time prevented me from doing a great deal of travel except of course while looking for work, which task was not at all difficult in those days as there were many more jobs than people seeking work.

So most of my desire to travel was heavily postponed to my anticipated retirement. Part of the desire to see as much of Australia as I could with Jenny after we had both retired, was to ensure that as we travelled we would visit every place I had visited during my early years so that Jenny could see the places where I had either spent time or worked.

Jenny expressed a great deal of interest in these plans and so we resolved that we would continue to plan until such time as we actually reached our retirement. Jenny who was always a person full of vitality started to show signs of slowing down slightly, coupled with an indication that some unusual uncertainty was creeping into her demeanour followed soon after by some signs of problems with her balance. In spite of the relatively short period of time since having arrived and settled in the area we were extremely fortunate to have Dr David Barker both as our Dr and our friend.

Initially the opinions from both David Barker and also my impressions suggested that Jenny may have been overdoing a lot of her activities, after all she was in full-time employment at the practice which fact she never treated lightly. After working hours she was never still with looking after me, the house, the garden all of which quite justifiably should have resulted in her being tired.

After a number of attempts on the part of David, he decided to refer Jenny to a physician, by the name of Dr Twomey, (since retired), who diagnosed Jenny as suffering from a condition known as Addison's disease which was evidenced by a very low output of cortisol from the adrenal gland.

Dr Twomey expressed the view that this condition was brought about by an excessive use of prednisone during long periods of treatment of her atopic eczema which in turn suppressed the output of cortisol from her own adrenal glands.

Dr Twomey recommended that Jenny should be given a maintenance dose of hydrocortisone and warned that if Jenny was ever showing any sort of illness, that the dosage should be significantly increased.

Here I must shamefully recall, that my attitude was quite deplorable when I now realise that I never made a conscious effort to ease Jenny's burden by at least attempting to take over some of her activities even though I knew that she would resist anything which would indicate that I found her efforts to be inadequate.

Things that were found to be amusing episodes, such as me sitting in my recliner chair watching television, lifting my legs so that Jenny could get under them with the vacuum subsequently turned into an embarrassment.

Looking back now I feel ashamed and I am quite convinced that the changes I had to undergo during subsequent times whilst caring for Jenny was hopefully adequate punishment for my insensitive behaviour. It provided me with greater appreciation of the myriads of people who hold down full-time jobs and coming home after a day's work, instead of sitting in front of the television or reading the newspaper, are further occupied with attending to the many things that needed to be done around the house.

So, as you read this if you can see yourself as I saw myself, I hope for your sake you will not be paying the same price. For the time, we noted very few if any changes until one Saturday while Jenny was working in the garden cleaning the water feature I heard a bit of a commotion and on going out into the garden I found that Jenny had fallen into the water feature which luckily was a very shallow pond and Jenny was able to get out of it saying that she was feeling foolish and that there was really nothing wrong.

On questioning her I became quite convinced that it was a loss of balance which caused the problem because such episodes were being noted with increased frequency but not necessarily resulting in a fall.

On discussing this episode with David, he suggested that he should refer Jenny to a specialist in Sydney who investigates such things as balance problems principally due to the problem in the middle ear.

By this time Jenny had formed a very firm friendship with our neighbour Maxine, who moved into her unit across the way from ours soon after we took up residence in our townhouse.

Jenny with Maxine, our "Guardian Angel".

An appointment was made and thanks to our dear neighbour Jenny was driven to Sydney and returned late that afternoon. That situation continued to recur.

At this time Jenny was still quite capable of driving her own car however considering the distance to Sydney and the likelihood of heavy traffic and taking into account her tendency to be off-balance it was considered unwise for her to undertake the trip unaccompanied let alone to drive.

On their return from Sydney, Jenny told me all about the day's events which unfortunately turned out to be without any definite result and necessitated yet another trip to Sydney for the specialist to conduct some further tests.

A further appointment was made and thanks to our dear neighbour Jenny was once again driven to Sydney and returned late that afternoon.

It was very disappointing that all these efforts and all the tests failed to reveal any reason for Jenny to have been affected by losses of balance albeit to us they were very real indeed and certainly could not have been something Jenny could have imagined. I doubt that it would serve a purpose for me to go on providing a blow by blow account of many such subsequent trips to specialists most of whom were in Sydney but also one in Canberra all of which revealed nothing which would suggest that Jenny's problems were able to be pinned down to any specific cause.

These attempts at a diagnosis occurred over a period of well over 12 months during which time Jenny's symptoms became more pronounced, with the losses of balance occurring more and more frequently but assuming a character more like a stagger.

During these times I did have opportunities to talk to people who were either afflicted by similar conditions or whose relatives were seeking to be diagnosed and it became fairly quickly obvious that Jenny's example of the time taken to finally be favoured with the diagnosis was by no means anything out of the ordinary.

My main concern as a result of all this was that by this time Jenny had commenced to become convinced that she was being a hypochondriac and it did not take much empathy to understand that a person would feel this way having attended a series of highly rated specialists using the very latest scans and tests were unable to find anything wrong.

Jenny had to surrender her driver's licence as, by now she was quite incapable of driving her car.

All of this in combination had convinced David that Jenny should consult a neurologist and as such a specialist was available at the Wollongong hospital, he recommended that an appointment be made to consult him. At this time this specialist was an employee of the hospital but was available for private consultations however, when the appointment was made and almost due to be attended something came up rendering the Dr unable to keep the appointment and it was agreed that a further attempt for an appointment be made.

At about the same time I was travelling from Canberra back home to Blackbutt after seeing some clients, I called on one of our clients at a service station in Goulburn who was quite shocked at the news of Jenny's state of health and indicated to me that he was a personal friend of a neurologist in Sydney who was a professor located in the Prince of Wales Hospital.

I was given the name and telephone number and when I got back to Blackbutt I telephoned and spoke to the secretary who in turn asked me to hang on and see if she could get the Dr to have a word.

When I told him who suggested I contact him he immediately asked for some more details which I gave him and he said he was treating someone with very similar symptoms and suggested for Jenny to be brought to him as soon as possible.

An appointment was made and this time I intended to take Jenny myself, partly to relieve our neighbour but also because I wanted to be present when the specialist examined Jenny.

It turned out that the man we were to see was in fact Professor Colebatch, the head of neurological research at the hospital and we could have faith in his pronouncements and after a relatively short interview and examination of Jenny, Professor Colebatch gave a provisional diagnosis unfortunately in the coldest possible terms as;

"*You have Multiple Systems Atrophy, there is no cure and no treatment!*".

On my request, the name of the ailment was written on back of a card and the word provisional preceded the word diagnosis which of course meant that the diagnosis was subject to confirmation.

This was in the middle of the afternoon on a Friday as I recall which meant that we were now facing the trip back to Blackbutt and we would be travelling out of Sydney at the height of peak hour traffic.

Thinking back now I am quite convinced in my own mind that Jenny did not understand any of what went on in the specialist's office but I do remember that it was one of the worst trips I've ever had to make of any distance with those words of the diagnosis still clearly ringing in my ears.

When subsequently reading up on multiple system atrophy, (MSA) every description of the diagnosis was virtually word for word repeating the words of the professor.

The difference of course was that reading those words whilst certainly conveying the same meaning to the reader, hopefully when passed on to the patient that any subsequent repetition of those words could contain some softening content no matter how vague.

We talk about "bedside manner" the purpose of which is normally, to provide home possibility of a light at the end of the tunnel so far as the terminal nature of the condition is concerned or some vague possibility of therapy which could soften the determination that there is no treatment.

After all as much as one may regard as absolutely important to be made aware of the extent of and the eventual anticipated outcome of any condition if for no other reason but the state of mind of the patient at the time of receiving such a pronouncement which certainly does not leave any room for any doubt that you are being told to go home and wait to die.

Both pronouncements could have had some softening additions, such as; "currently MSA is incurable" or "as yet there is no treatment" as well as the possibility of adding advice that "there may be some therapeutic measures which could be worthwhile investigating". At least there could have been the provision of a slightly open door through which a slight ray of hope could be perceived.

There is of course another impediment discouraging anything other than pronouncing cold hard facts in the delivery of the diagnosis and that is the constant and growing fear experienced by most professions brought about by the ever increasing number of people who will seek to make a lucrative living out of court actions for damages. I will refrain from expressing my true feelings on that subject.

As the diagnosis was of a provisional nature, there was an obvious need to obtain a second opinion and, having recalled meeting Dr Colin Andrews, a neurologist some years ago in Canberra I made an appointment to take Jenny to him for an opinion.

Colin Andrews didn't agree with the provisional diagnosis but instead expressed the view that Jenny was at the beginning of Parkinson's and prescribed "kinson" as the appropriate medication and suggested that we should telephone for a further appointment in about one month.

In the meantime, the professor himself was taking steps to confirm his provisional diagnosis.

The next period will in the main cover in some detail the period immediately after the pronouncement of the diagnosis. This began when we had a call from the professor's secretary asking to bring Jenny to Sydney to discuss the results of the autonomic function test requested after our last consultation. Following that meeting the professor subsequently confirmed his diagnosis.

A further appointment was made for three months down the track for Jenny to return and see the professor.

Looking back to before this final result, and considering the number of tests performed following several trips to Sydney resulting in no conclusions, this entire process unfortunately commenced to indicate to Jenny that there was really nothing wrong with her and that a lot of what was happening to her happened in her imagination.

If we are to recognise the importance of attitude, or perhaps more particularly the importance of a positive mental attitude, these procedures over an extended period of time were certainly capable of commencing to turn, what was always a highly positive mental attitude to a negative one.

It would be unfortunate if these factors were not being taken into account or indeed, if they are not being taken into account now, considering the added experience over time. These seemingly endless hopeful excursions, instead of succeeding in a clear diagnosis, are surely likely to be causing more of a problem than a benefit.

"FINDING THE WAY"

*F*inding the way is certainly an apt heading for the period immediately following the diagnosis.

Those who are about to become carers of a person awaiting diagnosis, will find themselves in a confusing state of mind which gradually increases in its level of confusion as test after test, scan after scan and numerous consultations and examinations appear to be without any conclusion and suddenly, out of that haze, the diagnosis happens.

A period of time when, after having commenced to find the way, there is a time of considerable shock and realisation of the changes gradually occurring in just about every phase of emotion and behaviour all of which requires attention which at this time not only involves assistance for the patient but in our case certainly also for the carer.

These changes I mention are not only physical and emotional but also necessarily practical inasmuch as the realisation that as this disease progresses, some quite substantial and gradual changes to the very way of life to which one becomes accustomed must be made. Having lived in a house which has to some extent been fashioned to your liking, you are now asked to examine it from the point of view of safety and other important considerations necessary when dealing with someone disabled and the actions recommended and needing to be taken to achieve the desired end.

So in describing the content of this period it covers a myriad of changes which gradually become obvious over time and are dictated in the main by the patient's condition and of course the type of residence occupied because not one place requires exactly the same modifications.

Unquestionably one of the most difficult changes which the patient needs to confront or is confronted by (and this was particularly real in Jenny's case), was to face up to the realisation that her belief in her status as part of us as a family namely the one who provides the care and does all the many things which made our life complete, to be gradually taken from her. To find her relationship to me to have necessarily changed with me taking over doing so much of what she loved doing. Such a realisation is devastating to a person who had always taken a great deal of pride in her abilities as a brilliant cook, who was always able to be quite faultless in every aspect of what we call domestic duties, resulting in the house always being spotless and yet not necessarily with nothing out of place because we both believed that a home whilst necessarily clean and orderly needed to show signs of being lived in.

As well as that for the 14 year period before her illness, she very capably managed our small but busy accounting practice.

I wasn't able to look far enough ahead in those days because of my lack of knowledge and understanding of how things may develop but I did anticipate that when it became necessary to have some outside help mainly to take some of these duties off my schedule because I commenced to need to give more time to the actual care of Jenny, that she would almost fiercely resent being suddenly faced with strangers coming into the house attempting to do the things she did so well.

The effect on me of all these changes was also predictably difficult to accept with the only saving grace being that whilst Jenny was becoming completely defenceless in an emotional sense as well as physical, I thank God that I managed to remain relatively strong and accept this new and strange burden for the sake of both of us.

So finding the way from having been told the diagnosis, the confusion didn't disappear, indeed it changed from asking yourself the question "I wonder what is wrong with her?" To an entirely new question being "Now I know what's wrong with her but what is causing it?"

Having been in professional practice for a number of years, having experienced some four years of involvement in politics and having been at the head of two national organisations, how come I was so helpless?

Anyone in a similar situation would be certain to experience an urge to find an answer to this relentless disease however, common sense will dictate that if we consider the number of people who suffer from MSA and an incredibly high number of other neurological conditions, most of which to a greater or lesser extent are subject to research by some of the most eminent brains in the world, it is best if we as carers and lay people, do as well as we can with what we have by way of therapies and the optimum quality and dedication of care and leave the answer to the questions; "why" and "what" to the professional researchers and assist with the funding of research as well as we are able.

I think if I have to describe my emotions at this time, on the one hand I felt some relief that the diagnosis was at last confirmed yet on the other hand the realisation of the devastating prospect of Jenny having MSA left me quite numb.

I am sure that the emotions of anyone else in a similar situation would also be attempting to come to terms with something relatively unknown.

It is even now difficult to come to terms with a situation where you are faced with some decisions to be made for the benefit of the person you are looking to support when not knowing what such decisions should be based on? I took my Jenny to David and after listening to our conversation Jenny commenced to understand that the condition she was diagnosed with was incurable.

There was no other knowledge or understanding she could have at that time simply because neither David nor I had much knowledge of MSA other than what little information we could gain from research of any material and any source available.

It was almost as if I sensed that we were facing a serious change, because I moved my office to one of the spare bedrooms at home and my Jenny was progressively transferring documents and had some friends help relocate what furniture we decided to take with us from the office.

By now it was obvious that retirement had become compulsory if I wanted to look after Jenny and so, I retired in October 2009.

I'm not sure why that should have been such an extraordinary decision when after all I recently turned 79 and under today's standards and conditions that would be regarded as far too late to retire.

I marvelled at Jenny's courage as she knew that certain things had to be done and did them with her usual aplomb sorting out and transferring documents to our "home office" but obviously, her demeanour had gradually changed commencing to show signs of being depressed.

She was subsequently referred to a clinical psychologist by David and treatment immediately commenced weekly. In my quest to gain as much understanding as possible I asked to sit in on the sessions which was readily agreed to.

The line of attack used by the psychologist was an attempt to convince Jenny that she should cease thinking about the past and try to recognise the fact that no matter how dire indications may have become there was clearly a period of time during which she could have some quality of life.

I am not sure that I was in agreement with this process as, given our patient's happy past on the one hand and her hopeless future on the other the proposal by the psychologist was a little difficult to justify.

These sessions were not altogether successful as whilst Jenny appeared to be very attentive and was cooperating, she did not show any interest in proceedings between the sessions.

As the psychologist gave Jenny some repetitive exercises to do between sessions, none of these were done and so, after discussions with David, the remaining sessions were cancelled.

I started to get to the point where each and every time something extraordinary happened for which I didn't have an immediate answer I felt I had to make time to find the answer.

The example of Jenny's sessions with the psychologist was such an example.

The question I asked myself was; "how come Jenny was paying obvious attention during the sessions and yet, was showing absolutely no interest in anything to do with the content of the sessions or the work she was given in between the sessions?"

As is often the case, common sense tends to provide the answers to such questions.

Given that we could safely anticipate Jenny's state of mind to be at a very low ebb when she was just coming to terms with the reality of the diagnosis, whilst her attention was captured by the obvious skill of presentation by the psychologist, she was naturally attentive but as soon as that capture vanished replaced by her very badly damaged state of mind, could we realistically expect her to just knuckle down and do her homework?

I have to stress that this practice of analysing everything in an effort to find answers must become a task almost as important as empathy, if the caring effort is to be successful.

To purely accept things as and when they happen without questioning the reason, will leave the job half done and therefore lacking an acceptable resolution.

There will be plenty of such examples which will tend to enable us to hone our skills in analysing what appears to be a mystery and we will find ourselves motivated by the anticipation that we will find the answer and use our new found knowledge to aid our patient.

I had to realise that now that we had been given a diagnosis it was perfectly obvious that there was a definite prospect that the level of necessary care given to the patient would be progressively increasing. I had very little knowledge of just what needed to be done and what arrangements were necessary to ensure Jenny's ongoing comfort.

I then recalled that when Jenny's late mother was diagnosed with cancer by her Dr in Goulburn, Jenny and I made the decision to bring her to Canberra to our home and to care for her as best we could there.

We were familiar with palliative care and when Jenny's mother was placed under palliative care we had a nurse visit each day and their advice to assist Jenny and I turned out to be invaluable.

This then gave me a foundation on which to build as well as having some friends who had some experience and as some of them were in nursing I was given a reasonable start.

By now there were some indications of Jenny's condition slowly becoming more and more obvious, with her coordination worsening, her walking commenced to become quite hesitant and unreliable and there was also an indication of the faults in Jenny's speech becoming slightly less intermittent and a little more slurry, all of which was indicative of a need for some more specific specialised knowledge and arrangements to be made for whatever therapy was available.

I began to realise that what was to become the greatest learning curve I could possibly have experienced had commenced, and I knew that whatever progress I was likely to make was for the benefit of my wife and patient.

Looking back now my anticipated level of learning curve then has become reminiscent of the size of a bright rainbow but without the attraction of a great deal of colour.

It was indeed fortunate that this was a period where the real level of caring had not yet become necessary and therefore as a carer I had plenty of time to think, analyse, research, question as well as read whatever information was and is available. For any carer the use of this time to ensure as much as possible that their knowledge, understanding and skill levels will be up to scratch when the real action begins is invaluable.

All these things should become a habit for those who care for others be they loved ones or complete strangers. They will remain carers because theirs is a calling and not a job and the need to care becomes a part of their very nature and character adding to the number already there and active, making the lot of aged and dying people somewhat more acceptable.

As this part of the book discusses the outset of the care of the patient, it is the most important part of the book simply because whatever actions are taken and in whatever manner they are delivered will likely form the basis of the service.

"Empathy"

*L*ooking back at the experience gained during my care of Jenny, let me say that empathy, its knowledge and understanding and subsequent practice should be regarded as the foundation of the relationship which will likely exist between carer and patient.

Similarly to having understood and appreciated empathy, knowledge and understanding of anything and everything which is likely to affect the patient under your care will assist in ensuring that the actions taken by you will be the right ones and not only will not cause your patient any distress but rather it will be obvious that whatever action was taken, was done to prevent any distress.

My knowledge and understanding of MSA as a disease had grown somewhat and the most important part of this period to me had to be the opportunity to muster up all the ability I had to practice empathy to in turn help me to almost constantly walk in Jenny's shoes, think the way she thought and feel the way she felt.

We have learned when we were growing up having been part of a devoutly religious family, the reference in the Bible to "do unto others as you would have them do unto you".

Whilst that passage from Scriptures does teach us to some extent to learn to understand and appreciate the feelings of others, it was more part of the instruction that we must never do anything to anyone which we would not accept being done to us.

Empathy is different. It means that you virtually become the other person in every respect, even down to feeling the same temperature as they do, hearing every sound the way they hear it, reacting to every happening as they do. To empathise means that you are capable of understanding and appreciating everything that is happening and is being experienced by your patient.

If I can appreciate this as a carer and it will take some considerable time to achieve it, I will feel uncomfortable if the patient feels uncomfortable, I will enjoy what they enjoy but I can hate what they hate. To empathise means that you are capable of understanding and appreciating everything that is happening and is being experienced by your patient.

I now believe that every person involved in caring of any kind but in particular those caring for a loved one should diligently learn every facet of empathy and not before then will they be true carers. We mustn't pretend that empathy only exists on one level or that it is necessarily applicable in one-on-one situations.

I believe that the ability to empathise is much more dependent on character and can be regarded as a behaviour.

Over the years I have discussed this subject with many people and it has been the general consensus that there are people quite high in number to whom the practice of empathy is a perfectly natural human behaviour.

I would use Maxine as a typical example of this because, for all the years I've known her and associated with her I perceived this behavioural talent in Maxine right from the beginning and what is more important, I have noticed it growing as her relationship with her granddaughters developed.

We could say that one of the most important components of empathy is the ability to anticipate as long as the anticipation is used and practised in terms of others and not of yourself.

Jenny was, over the years very much a part of these discussions and her understanding of at least some of the matters we discussed did help me greatly when I commenced nursing her.

I remember Jenny being highly amused once when I made the comment that I didn't bother to try and use empathy on Chloe for obvious reasons. To try it would have assumed that the behaviour of a cat is predictable.

It should also be recognised that empathy can be practised on different levels and it would certainly be my opinion that the highest level of empathy would be from one person to the other where there is love involved.

I think that there is a certain amount of logic in that comment because when you love a person and when you are together as a couple for example, some of that anticipation of your partner's behaviour commences and grows as does the relationship.

We could say therefore that the longer a relationship exists and flourishes, the easier and more natural the practice of empathy becomes.

I appear to be belabouring the point but I do feel that there is nothing more important or effective in caring than the true practice of empathy.

I also believe that even complete strangers who will become carers and who are blessed with the ability to become emotionally involved with the person for whom the care is provided, will need to understand that the practice of empathy is an integral part of caring for others. There will be plenty who will criticise the statement in the above paragraph where I claim that becoming emotionally involved with a complete stranger under your care is a blessing, which probably does go against some institutional advice given from time to time to carers warning them not to become emotionally involved because it can interfere with work routine.

The intent behind the contents of this book is primarily for the benefit of carers but of course ultimately, the true benefits flow on to the patient under care, and there is no doubt that providing genuine care more as a calling than as a job, can and likely will cause some pain to the carer but that is what makes caring a sacrifice.

Because this part discusses the outset of the care of the patient, it is the most important part of the book simply because whatever actions are taken and in whatever manner they are delivered will likely form the basis of the service.

Accordingly let me say that empathy, its knowledge and understanding and subsequent practice should be regarded as the foundation of the relationship which will likely exist between carer and patient.

"SPEECH THERAPY AND CHOKING"

*I*t was at about this time David recommended that we should engage the services of a speech therapist for Jenny.

As at this stage Jenny's speech was not yet a real problem and I recall wondering why this recommendation needed to be made, but steps to arrange such sessions were taken.

This was a typical example of using time available to obtain as much information as possible in order to be able to face occurrences as and when they happened. What I learned about Jenny's speech and the need to consult a speech therapist was that we were really talking about one of the more serious perils faced by sufferers of MSA.

I learned that the gradual deterioration of speech may have been and probably was an early sign of the gradual failure (atrophy) of coordination.

We know of course that without coordination we would become physically crippled because in reality the left hand will not react to the right hand, the right foot not react as expected to the left and from that commences our understanding of how we can reach our own conclusions and work out what part of our body is likely to react adversely and subsequently fail if we completely lost coordination.

One of the major contributors to our ability to speak clearly is the role played by our tongue because without its movements and actions, whether our tongue was touching the roof of our mouth or extend out and touch our lips and many other essential movements, our speech would become garbled and we would no longer be understood.

We were advised that there was speech therapy available from the local hospital and so David prepared a referral and an appointment was soon made to attend a speech therapy session.

As had been and would continue to be my intention I attended each of these sessions which enabled me not only to carefully observe and follow the actions of the therapist but also the reactions evident in Jenny. as it did happen at the outset of the sessions with a psychologist, and as I had subsequently analysed why that program wasn't a success, it was discontinued.

The first speech therapy session consisted mainly of Jenny's speech being tested no doubt to determine which part of her speech contained some anomalies.

Katie, the therapist would read something in short sentences and Jenny was asked to repeat what she heard. Jenny was also expected to read and was being observed from the point of view of whether she was aware when she was making a mistake. At the end of the first session Jenny was given some sheets of paper with single letters and combinations of more than one letter and was asked to read them and repeat them for homework so that by the time she came for the second session they would all be familiar to her.

It occurred to me at this time that Jenny was being tested by the speech therapist with a view to ascertaining whether Jenny's cognizance levels would still be able to develop habits through repetition.

I formed the view that any habits which can be developed during the early stages of the condition could aid in prolonging the patient's capabilities or hopefully retard the progress of the atrophy.

Subsequent attempts albeit few and very brief, did not indicate that such action was likely to be beneficial.

In fact, the exercise involving single and dual letters was more to test the way in which the various letters Jenny read and pronounced were formed and would in turn indicate whether or not the tongue was doing its job.

This again put me in mind of the episode involving the psychologist and I remember being quite determined that whatever Katie asked Jenny to do by way of "home work" I would ensure she would certainly do and to ensure that, I did each of the tasks with her.

It is important to appreciate that there was nothing wrong with Jenny's brain at this time nor was it anticipated to fail at any time and so in this situation the importance of empathy once again came to the fore and I was pleased to note that the therapist was well aware of this.

The next number of sessions commenced to concentrate not only on speech but Jenny was asked to eat and drink small amounts of food and liquid and I noted that Katie had her finger on Jenny's throat monitoring the swallowing action.

This certainly was a period where I needed to make a special effort to read up on what I was experiencing which subsequently helped me a great deal to anticipate what was likely to happen to Jenny as she developed the problem with her speech and particularly with swallowing.

I learned a considerable amount by reading, talking to others and following the growing number of entries on Facebook which by now commenced to be used by sufferers and carers alike mainly asking questions though some were also providing answers. The connection between coordination, speech, and the actions involved in eating, drinking and swallowing started to become clear to me.

I believe the reason David recommended the use of speech therapy was because all those other related actions come under the area of a speech therapist's expertise.

I learned much later that whilst the loss of coordination was a very serious peril, the event involving swallowing was to prove far more serious, understanding of course that the loss of coordination had to be recognised as the source of the problem and would in many instances bring about the end of the battle.

At this stage Jenny was still a long way from experiencing serious episodes with swallowing, but it is very important in the process of caring for terminally ill patients to be able to look ahead and have as much information as is available to anticipate the problem and do what can be done to keep the patient comfortable.

There will be numerous further references to these problems as and when they occurred and unlike my lack of knowledge and understanding of these things when I was caring for Jenny, the readers of this book depending on the length of time they have been involved in caring, this will not all be new.

In my case as in the case of many other carers of my vintage looking ahead generally meant using some logic and some luck whereas today it is a much improved scenario because any carer who commenced after some training and takes notice of carers who are experienced will have plenty of guidance to be able to anticipate the progression of at least some of the debilitating conditions.

For that reason I have decided to include here as much as I am aware regarding the problem of choking which of course became obvious when I watched the therapist observing Jenny eating and drinking and feeling by placing her finger on the outside of the front of Jenny's neck gauging the swallowing movement.

As coordination continued to fail all actions involved in eating and drinking as well as speaking gradually worsened. The failure of speech on its own is of course a problem but not a danger.

Where the failure of coordination creates the danger is that what is an automatic process when we take food, chew it and at the right time swallow it which in turn ensures that the substance taken and chewed will descend using the correct exit from the mouth if I may use such a mundane expression.

Where it descends is the critical point because of the two ways it can go, one takes it to the stomach which is correct but the other which is used to facilitate your breathing leads to the lungs.

Foreign matter other than the air you breathe entering the lungs in any volume can and is likely to cause lung infection which can and frequently does lead to pneumonia.

When this happens in small amounts it usually causes a cough and we normally refer to it as "having gone down the wrong way".

Quite often small amounts are retrieved by coughing however there will always be times when that will not happen.

Whenever there has been a serious case of aspiration that means that the amount of food that found its way to the lungs was excessive and the reaction will as a rule be choking and if the blocking substance is not retrieved, an infection of the lung is likely to follow.

It is not only large amounts of foreign matter that can bring about an infection. This also applies to smaller amounts which, not necessarily by themselves cause a problem however their accumulation can have the same result as one single large aspiration.

Quite regardless of whether we are talking about our patient being in a nursing home or similar institution or, still at home, neither would be equipped to handle a case of lung infection and so an ambulance needs to be arranged for the patient to be taken to the nearest hospital.

The most common treatment is usually an antibiotic and once the infection is clear, the ambulance is once again called and the patient returned.

I am sure we are all aware of the constant shortage of beds at all hospitals and consequently any upsurge of hospitalisation as a result of choking is not appreciated by the hospital system and patients so admitted are normally quickly attended to sometimes by administering antibiotics intravenously, and returned to their home base.

Unfortunately, because these episodes are brought about by the progressive failure of coordination, they will almost certainly re-occur and are likely to do so at an increased rate of frequency.

Each time this occurs your patient will be visibly stressed, not wanting to be taken to hospital again and of course the time will come when the Drs at the hospital will become increasingly reluctant to administer more antibiotics and so it is likely that the carer will be advised that the point is being reached where any further infections will likely accelerate the progression of the disease.

It is appropriate at this stage to make mention of an alternative which can be and is often offered and that is a change to artificial means of feeding the patient the most common of these being by the use of a tube.

By this time it is highly likely that your patient has an Advance Care Plan which, depending on the wishes of your patient either allows such an alternative or refuses it.

Many of the discussions I had with people involved in end-of-life care indicate that being fed by artificial means does not necessarily prevent the likelihood of choking.

It is true to say that the process of artificial feeding itself will not bring about choking but the possibility of the regurgitation of food remains likely to occur which in turn can bring about choking.

I am including all this at this time because it is important to be aware of these things likely to be occurring as indeed, was the case with Jenny and so it will be gone into with further detail later in the book.

"*Self Hypnosis*"

*J*enny had a tendency to have spasms and speaking incoherently while sleeping and these had been a feature of her ongoing condition for some considerable time.

These tended to occur at various times during the night and most times the spasms and speaking occurred simultaneously.

The spasms as a rule involve all four limbs, at times representing almost violent movements.

On occasions speaking occurred with or without small spasms but with an infinite variety of durations as well as containing various volumes of voice up to the level of screaming and though the content is incoherent, there were times when Jenny was either crying or laughing.

This would tend to suggest that she was dreaming, but when awakened she did not recall anything.

This condition had endured over 12 months and there appeared to be no change in the frequency nor was there any pattern to the manner in which they were manifested.

Early one morning at about 6.30, I observed Jenny literally leaping out of bed and donning her dressing gown. The use of the word "leaping" has some significance, inasmuch that each time she had occasion to get out of bed it took some minutes for her to reach the sitting position on the edge of the bed, where she usually remained for a few moments before reaching for the rod fixed across the wardrobe door, which she used as an aid to stand up.

At this time Jenny was no longer able to stand unsupported and she was unable to walk without assistance.

On this particular occasion, having leapt out of bed donning her dressing gown, she stood unaided.

When I asked what she was doing she replied; "I don't want the food to get cold".

I then asked, "What food?"

To which she replied; "The food on the tray". and with that Jenny pointed at the bed and said with a raised voice "Where is the tray? – It was there a minute ago."

I then took Jenny by the arm and helped her back to bed.

The significant thing was that what she said she said with complete clarity, without any sign of a slur in her speech which by now had become quite slurred and difficult to understand.

Two days later I was awakened by the light having been turned on outside our bedroom on the top landing of the stair well and I saw Jenny, again standing unaided, at the head of the stairs.

When challenged, she said, again with a voice as clear as a bell, "I am looking for Patrick who has gone downstairs to get a sandwich".

My regret is that I was not awake to have witnessed her progress from the bed to the landing.

When I escorted and assisted her back to bed, she was in need of assistance but by then she was no longer dreaming and seemed to be wondering what had occurred.

By the way, Patrick is a friend who used to sometimes visit us.

After these events occurred, I waited before completing these notes hoping that a similar event would re-occur.

Just what significance if any these occurrences may have, I am not qualified to say however, I cannot help but ask; "Is a dream coupled with sleep walking a type of trance?"

If the answer to this was "Yes!" then the next question is "Would there be some benefit in consulting a qualified clinical hypnotherapist?"

After all, if in whatever state Jenny was at the time had been able to reproduce her pre MSA condition, would it be possible that a hypnotic trance could achieve a similar result?

I made contact with a number of hypnotherapists on the internet and managed to get more information all of which was of great help in my attempt to be informed.

Clearly, any use of hypnotherapy is not to be taken as a possibility of a cure but rather its use in suggestions aimed at Jenny's ability to recognise the importance of the present as an interim period during which if successful, she may be assured of a more prolonged period of an acceptable quality of life.

Without doubt it is appropriate to mention that at the time Jenny commenced sessions with the psychologist there was a very similar concept involving a focus on the period of life remaining. However there was also the accent on completely ignoring the past which I felt may have been inappropriate.

If hypnotherapy proved fruitful, it would assist greatly in the development of a state of mind which could accept and visualise a positive period which would hopefully result in enough motivation to induce Jenny to fully cooperate in any program devised for her benefit.

Sometime later I engaged a person who came every Thursday morning for five hours to do the washing and the ironing and to keep Jenny company while I played a round of golf.

As well as all that, I had Jeanette, Jenny's sister who came at midday each Tuesday to stay with Jenny while I had another round of golf this time with our Dr, David Barker.

I used to find it difficult to carry on with my notes due to fatigue at the end of each day.

These new arrangements enabled me to spend evening time on my notes as well as do a lot more reading which for the time being was to gather information on various forms of hypnotherapy.

Most of us have some awareness of the power of the mind and we are also likely aware of the effect on general performance by a person who has a negative frame of mind as against one whose mind is governed by positive influences.

As I read volumes of facts and information on the ability of hypnotherapists to influence and considerably improve the mind of persons by positive suggestions, I could not help but conclude that the same regrettably is true in reverse. In other words; negative suggestions can have a negative influence on the person's demeanour.

I go back to Jenny's period of tests and examinations which, until the diagnosis finally came, must have played havoc with Jenny's thinking indeed, quite often Jenny said that she was becoming convinced that there was nothing the matter with her and it was all in her head!

Then came the diagnosis, which first stated that she had an incurable condition and then she was told that there was no treatment.

I know that Jenny did not remember either part of the diagnosis but I am convinced that it was retained in her subconscious and, to make matters worse it was reinforced each time she heard me repeat those words to others and every time she read something I downloaded from the Internet!

I had decided that nothing should be kept from her and so it was not long before she felt justified in the attitude of "what's the use" whenever she was presented with an idea by way of some attempt for therapy.

I will never forgive myself for not having taken steps earlier to somehow counteract the terrible effect that this negativity had on Jenny.

To me, it stands to reason that if a person is told that his or her condition is terminal and that there is no treatment, the pace of the degeneration which is part of the condition could accelerate because of the negative frame of mind.

If there is a way in which the rate of degeneration can be reduced, it would in turn increase the period of a reasonable quality of life. If that can be achieved, should that not be regarded as treatment?

During my research I encountered the Goulding Institute, the organisation which markets a hypnotic program named "Sleep Talk", developed by the principal of the institute, Joane Goulding.

I purchased the book and after studying it I contacted Joane by email and put to her the proposition that whilst "Sleep Talk" was designed to be used by parents especially trained to deliver specially composed hypnotic messages to troubled children, designed to counteract and reverse negative influences, it could be re-developed to be used for adult patients for the same end result.

This suggestion must have struck a chord as I received a reply within 24 hours and was told that my proposal was already under consideration and that we should meet and have discussions.

Joane was due to travel to Sydney to address a symposium of clinical hypnotherapists on Saturday the 28th April 2012 and she proposed to arrive on the 27th to have the opportunity to meet with me. This was entered into my diary and was arranged to take place.

In the meantime she recommended that I should obtain four books;

"Keys to the mind" (Learn how to Hypnotise), by the well known therapist, Nathan Thomas.

"The Brain that Fixes Itself" by Norman Doidge MD

"Evolve your Brain" by Dr Joe Dispensa and

"Quantum Self Hypnosis" by Jo Ana Starr PHD

Because Paul, my son, very generously gave me a Kindle as a birthday present I was able to obtain these books from Amazon and had them all within one day electronically.

I did not get much sleep for the next few days!

The information I gained by reading and absorbing some of the content of these volumes was stunning.

The first thing I did was to read the first book in two days and immediately set about to work on commencing a program of hypnosis.

I was advised that, to begin with, I should not try to hypnotise Jenny one to one because there would likely to be a disturbing distraction and perhaps reluctance on her part to accept commands which may be part of the procedure.

I commenced to script some sessions and, having purchased a digital recorder, I recorded them in front of my computer, using itunes to provide some soft classical background music.

I emailed the scripts to Joane who approved them and thus commenced the first hypnotic sessions which by the day before my meeting with Joane had been going, at least twice daily with some success.

No one should suggest that an 82 year old brain cannot be capable of absorbing and retaining unfamiliar information.

I owe a great deal of gratitude in particular to Joane Goulding, the principal of the Goulding Institute whose very active help especially during the commencing stages of my project was certainly responsible for the relatively short time it took me to reach the stage of gaining the ability and understanding the process of self hypnosis.

I also owe a great deal of gratitude to my son Paul who, notwithstanding the considerable distance from where he lives, always managed to come to my aid and assist with the necessary equipment and software to make the preparation of these self hypnosis sessions a reality.

There were a great many questions which I needed to have answered and Joane was certainly instrumental in giving me the answers.

My first question of whether the process posed any danger was very quickly answered by the statement that because the program is or will be developed to use the successful sequence of positive messages by repetition, all of which will ultimately benefit the recipient there is no danger in the procedure.

Clearly, the program should commence as originally intended, targeting Jenny's negative frame of mind recognising that the two main thoughts on Jenny's mind were those contained in the diagnosis.

Because the time which had elapsed since the diagnosis that negative state of mind would indeed have become almost hard-wired in the brain and the aim therefore had to be to replace that negative state of mind with one which was positive.

The comment was made during the discussion that had this process been implemented soon after the diagnosis the progress would have been easier and that thought should be kept in mind if a similar undertaking is ever to be used for other people suffering a neurological illness.

In Jenny's case there was yet another question relating to the fact that Jenny was progressively developing a serious short term memory problem however Joane pointed out that this process is directed at the sub conscious mind.

It was stressed that it is absolutely essential that I maintained a completely positive frame of mind and that in day to day conversation the content must at all times be forward looking and positive.

The program involves the writing of the script for each session, then recording it with background music, editing the recording and re-recording it if any problem is detected. Naturally if I was using professional equipment the editing process would be somewhat quicker.

This was discussed with Paul and he installed some additional software on my computer which enabled me to record the script into this special program and that same software was able to accept musical tracks as a background to the script recorded and then play back the complete recording which could then be tested and if any changes needed to be made this was able to be done.

I was most grateful for this because it significantly reduced the time involved in the whole procedure which in turn enabled me to record a number of sessions thus introduce a considerable amount of variety into the whole program.

Jenny would have a session mid morning after her exercises, again in mid afternoon and after having retired for the night, a further session using a pillow speaker.

Joane warned and I think I was already aware that this process would be a long one and that Jenny would continue to have periods of depression and would continue to express doubts for some time to come before any sign of a reversal of her state of mind would start to become obvious.

By halfway through May I had prepared and recorded some ten different scripts and I was pleased to note that there were times when Jenny actually asked to have an additional session as I think she enjoyed the relaxation which is part of the process as the recording commences.

From mid May I decided to introduce the concept of a dream which would begin after the hypnotic trance commenced.

There were three dreams at the start and each of them covered one of the problems which caused Jenny to become depressed or at times even angry.

The main one was her beloved garden and the presence then of Chris who came and attended to the garden once or twice a month. Jenny resented his presence because she considered that she should be the one looking after things.

The second dream was the laundry which, now that there were some signs of incontinence, necessitated my being involved not just to do the laundry but becoming involved in the more intimate activities brought about by that condition. Jenny was suffering acute embarrassment and nothing I said could change that.

The third was the kitchen which was always Jenny's exclusive domain and her skills in that department were never in question.

She resented my having taken over and that there was really nothing she could do about it.

The dreams simply gave her the experience of "being back" starting once again to do the many things she used to do, with the accent on regaining control of her life.

The day to day conversation followed through with soft discretion, when we talked about how she would go about doing this or that when in time she was once again able to do them.

After a while I received advice from Joane to discontinue the specific dreams:

> *"I've been thinking about this. The dream was a great start however now that she has started to become used to the process, you might consider to allow Jenny to determine where her dream goes herself. So use generalised statements rather than specific perhaps at this stage. Becoming too specific might actually restrict her thinking at the deep subconscious level.*
>
> *With the Sleep Talk suggestions, we used two words that allowed the mind to determine what they wanted it to mean. The 'WE' allowed for whoever the mind wanted it to be. And the 'IT' also allowed for anything that related to that word. So perhaps becoming 'generalised' in your suggestions might be more appropriate and of deeper meaning rather than trying to determine Jenny's thinking"*

Back to the drawing board!

Jenny was particularly keen on one session scripted specifically for positivity and relaxation and requested it frequently.

The realisation of the devastating prospect of Jenny having MSA left me quite numb and I was driven to do more to try to achieve better results for Jenny's everyday existence.

There were times, increasingly so, when I wished there were two of me. One to do the hypnotic process and the other to do the caring which, in itself is very close to a full time occupation.

To remain positive through all this was as great a challenge as I have ever faced! Prayer remained an essential part of whatever was left of each day because for this to succeed and for me to be there to see it succeed would not happen without God's help. Miracles are very hard to come by and yet, wherever we turn we see miracles and none greater than the power and versatility of the human brain, understanding and the harnessing of which can bring about many miracles.

"REHABILITATION SESSIONS"

The speech therapy sessions came to an end and I did ask the Katie if there was a possibility of them continuing and was advised that the hospital rehabilitation section had decided to offer a special concentrated series of sessions to patients referred by their Dr which would include speech therapy, physiotherapy, Occupational Health & Safety, all of which would be presented as a clinic with physical participation involved.

I asked Jenny if she would like to participate and whilst there was some hesitation on her part she agreed and David provided us with a referral.

This meant that Jenny would commence attending a series of sessions at the rate of three times a week, Monday Wednesday and Friday over a period of four weeks.

The commencement of therapy sessions convinced me that Jenny would require some walking aid because more often than not attending these therapy sessions at the hospital would require parking the car some distance away from the entrance of the rehabilitation section of the hospital and any distance was becoming excessive for Jenny even though I assisted her.

At home under my care

Assisting her was quite difficult because Jenny was barely 5 feet tall as against my 6 feet and 4 inches. For us to be able to walk arm in arm whilst quite delightful became somewhat of a balancing feat.

At this time I decided to purchase four walkers, one for each level of our residence and one to be in the car at all times in case I needed to take Jenny to various places from time to time.

They had four wheels and a seat which enables the patient to sit and rest if and when necessary. These walkers were an excellent means of transport for any disabled person however, most of these devices also carry with them some risks which can cause serious problems if adequate and appropriate training is not given.

Training at the outset and progressively increasing assistance and supervision as the patient loses more and more of their coordination and ability to walk were essential.

The walkers are equipped with a brake for each of the two rear wheels and these can be engaged spasmodically as needed by raising the lever which is attached to the main handle held by the patient whilst using the walker.

At the same time if those levers operating the brake are depressed downwards the brakes will be able to be locked which is essential each time it is necessary to stop for some reason and also if the patient desires to take advantage of the seat. In order to sit on the unit the patient either has to turn the walker around or alternatively walk around it themselves and if the brakes are not locked, as the patient goes to sit the back of the knees push the walker backwards which then makes it roll back and the patient could fall on their back risking serious injury.

These changes proved themselves to be very satisfactory for as long as Jenny's condition didn't deteriorate to a point where some additional safeguards became necessary.

Soon, it was time for Jenny to commence her concentrated rehab sessions at the hospital and so starting with the first session, I drove Jenny to the hospital taking the car right up to the entrance of the rehabilitation section, preparing the walker and placing it just inside the entrance getting Jenny to sit on the walker making sure the brakes were on and I told her that I would join her as soon as I managed to park the car.

This worked wonderfully well and it was one of many examples which proved to us that the walker was the answer and the capacity for Jenny to be able to sit when she was tired was a bonus.

I gained the impression that Jenny was showing interest in the sessions and thinking back I am convinced that changing from one discipline to another on each of the days helped to capture her interest.

Apart from the sessions of speech therapy Jenny enjoyed the exercises in the physiotherapy portion of the sessions but I felt in the main, that it was the fact, as in the case of Katie, that she and the physiotherapist related well which was pleasing to see.

I am afraid as was the case on previous occasions I could see some benefit for the entire period of the sessions but that soon ebbed once they were over.

During a brief talk with the physiotherapist she pointed out, quite rightly, that there appeared to be no discernible sign of any motivation in Jenny and she was of the view that without that none of the attempts in any of the disciplines would be likely to be effective.

Nothing is more likely to reinforce my assertion that any activity which can fill the patient's mind with anything of interest will, as a matter of logic replace, if only temporarily, the negative thoughts which tend to reside there.

No one who knew Jenny and who was in any way familiar with her then circumstances would have any reason to doubt the state of her mind following what she had been through.

The fact that Jenny was a naturally positive minded person whatever her situation, her current downward spiral into depression was very much against the norm and it would therefore be reasonable to assume that any attempts at motivation to find something that created an interest in her mind would be very beneficial.

Those of us who studied motivation would agree that by far the majority of us humans are more likely to be negative than positive otherwise many more people would claim to be successful.

So, to motivate an average person would, I believe require less of an effort than motivating one who has been literally forced into a completely negative frame of mind.

One of the simplest forms of motivation which we recognise is an individual's ability, or perhaps we should say an individual's possessing a reason, to look forward to something positive in the future.

We recognise this in sport where motivation can be gained from a negative point where the sports person has just experienced a failure, to then induce a change to a positive frame of mind would result from the encouragement that their next result will be bound to be better.

Let me use golf as an example where I have just completed the worst nine holes ever and when standing on the 10th tee I resolved to make the second nine better.

Naturally, the success of the result will depend essentially on the level of my dedication, because if there is no conviction behind the attempt a further failure is very much a possibility.

If we examined Jenny's state of mind from the point of view of motivation, the disturbing question which would arise is, what if anything has Jenny got to look forward to?

This is where we can talk about that interim period during which points of interest can be injected and even though this is likely to be a very short term result it is certainly better than a complete failure.

"THE CRUISE"

*I*n October 2011 Jenny's sister Jeanette and her husband Albert asked Jenny and I to join them on a cruise to New Zealand with the duration of fourteen days, departing Sydney Harbour and returning there fourteen days later.

Whilst this was a very interesting and inviting suggestion, it brought up many questions most of them in relation to Jenny's condition and speculations as to what extraordinary things can happen on a cruise with which a disabled person may not be able to cope.

At this particular time Jenny, whilst having difficulties with her speech, was still able to make herself understood.

Her walking unaided was practically non-existent.

The most important question of course was whether or not Jenny would like to go on the cruise and if, after departure she found herself disliking it, what then?

She certainly would not be able to swim back and her state of mind under such circumstances would be likely much worse than had she stayed on land.

These things were discussed with many people before a decision could be made, not the least of which was a person from the cruise line whose duty was more of a social director. She quoted some data which indicated that there was ample evidence that most disabled passengers loved being on the cruise, loved all the activities and there was nothing on record which indicated anything to the contrary.

We then needed to talk about Jenny's inability to walk unaided, and had to contact the cruise line to ask them whether or not it was acceptable for us to bring on board a walker as well as a wheelchair.

We were informed that when we arranged and completed the booking we should stipulate a disabled cabin which is considerably larger than the normal size cabins and both the items in question could be comfortably stored in the cabin without causing any risk or discomfort.

So far as Jenny's speech was concerned there was really not much that could be contemplated and I made the decision to resolve to be very guarded about with whom Jenny had the opportunity to communicate, especially during the early stages. As time passed I introduced Jenny to some strangers and just used my judgement whether or not she could handle them.

The one thing I found most appealing, was the fact that both the place of departure and the place of arrival were the same and so it was a matter of going on board, unpacking all that needed to be unpacked and not have to pack again until the final day.

The day of our departure arrived and all went very smoothly. When we arrived at our cabin I was indeed very surprised at the size of it and we had no difficulty at all fitting everything in including the walker and the wheelchair.

We met the attendant who was looking after us and found him to be a very congenial little gentleman from the Philippines and soon, within the first couple of days we realised that the cruise ship was almost entirely staffed by "little Filipinos".

As soon as we had settled into our cabin Jeanette came and suggested that we go to one of the lounges and have a drink, relax and get a feel for the atmosphere.

For those who have not yet been on board one of these giant cruise ships, they have several levels the bulk of them taken up by passenger cabins of various types and sizes. On the middle and upper levels there were several restaurants of various types including two formal dining rooms, and a number of lounges most of them complete with a bar and all very lavishly and comfortably furnished and if our vessel was any indication everything was done in good taste. Then came the upper decks with their swimming pools, outdoor entertainment areas as well as provisions for alfresco dining, in short, not much if anything was left to the imagination.

I was absolutely thrilled with the response I was perceiving from Jenny who seemed delighted to be joining Jeanette and Albert in one of the lounges. We joined them for drinks and I noted Jenny being very relaxed, even smiling at some of the passers-by.

There were some quite extraordinary changes evident in Jenny's behaviour, she wasn't looking to have a rest during the day, she enjoyed joining us having a choice of various eating places on board, and she ate exceptionally well.

We made it a habit to always have our evening meal in one of the two main dining rooms where we had a table for four reserved for us and we were looked after by two very nice waiters one being a man the other a woman both from the Philippines. They were exceptionally attentive to our needs and became very fond of Jenny and she was beaming every time we went up for our evening meal.

The vessel had two theatres and there was a show every evening during our journey and just to continue my amazement Jenny insisted on attending one each night.

Jenny was mostly in the wheelchair partly because she found the walker rather tiring on board and the other reason of course was that although it was a very smooth trip, there was some natural movement which she found a little hard to handle with the walker.

At the back of each of the two theatres there was a section reserved for the disabled and there was a space which accommodated the wheelchair and so Jenny had a perfect view of the stage and all this must have seemed like a special holiday to her.

I would not be truthful if I didn't admit that it certainly was a perfect holiday for me too because to some extent it was two weeks of respite not needing to do things like making beds and cleaning the room etcetera.

Naturally our companions were also a great help as Jenny's brother-in-law Albert very often took over the wheelchair leaving me free just to walk and Jeanette was also very helpful and attentive.

So whenever it was decided to go ashore to sight see or perhaps to do some shopping, when Albert decided to take over the wheelchair he only knew one speed and that was flat-out which made it rather difficult for me to keep pace. Fortunately this proved beneficial because Jeanette and Albert had been to New Zealand before and so Jeanette and I could fall behind to the speed of my limp due to a bad hip and Jeanette seemed to know where Albert was heading.

As it happened, Jenny and I only missed one opportunity to go ashore and to be perfectly honest, it was more my fault than hers because I didn't really feel like another run.

The surprises continued because Jenny showed great interest in going ashore looking at things, buying a few things and naturally the cameras were busy and in many respects things were a bit like our normal times.

It is not my intention nor would it serve any purpose to go through the day by day experience of the cruise but what is important for me to state that virtually from our arrival on the ship Jenny's mood visibly lifted.

She enjoyed the attention she was receiving from the staff, she loved the surroundings and the fact that there was constant movement and a great many people which would normally have frightened her but on the ship she blended in and was very much at home.

Let me take us back now to where I talked about that gap which existed between then and the inevitable, and there is no doubt that an opportunity like that cruise would be one which would provide not just a short-term benefit but in Jenny's case a full 14 day period during which she really did have a chance to forget all the depression all the fear and concern and had a short period of joyful time.

Added of course was that with all that activity she was worn to a frazzle by the time we got back to our cabin at night and as a result she slept right through the night and it was obviously a restful sleep because she was bright as a button next morning.

"Facing up to Reality"

*T*he title of this section is really directed at me as the carer because even though I was aware of what our future held more so from Jenny's point of view than mine, when the time came to consider what options we had to ease the pressure somewhat, I realised that I hadn't really completely faced the reality and finality of the situation.

When it became obvious that some outside help would be necessary, we were advised to get in touch with ACAT, the government funded organisation which has responsibility for assessing individual patients affected by a disease with declining conditions and as a result of that assessment make recommendations as to which of the government programs available would be appropriate for the patient in question.

It is appropriate to understand that, whilst we appreciate any efforts on the part of the government to make such services available, we must of course understand that encouraging disabled people to remain at home for as long as that is possible, is ultimately a major financial benefit to the government because of the relatively small amount of resources employed in an effort to help people remain independent at home compared to the cost of residential high care.

Naturally, there is a desire on the part of everyone to be allowed to remain in one's own home but in many if not most cases when the physical condition of a person deteriorates beyond where it can be adequately cared for at home, a move to a residential high care is unfortunately inevitable.

ACAT came and assessed Jenny with the result that she was found to require low care at that stage with advice that as and when her condition deteriorated further an additional assessment would be appropriate.

As is more often than not the case, the assessment is but the first step in these proceedings and it is not to be assumed that the recommendation made as a result of the assessment will immediately become a reality.

There are a limited number of organisations that are appointed by the government to provide carer services and because funding as always is limited, it is often some time before the service selected becomes available. In Jenny's case this was about 2½ months after the date of the assessment when the organisation, the Anglican Retirement Villages offered the service and after we had the details explained to us the service commenced.

At that stage the service only consisted of cleaning where a cleaner came once every two weeks to clean the house and it was probably apt to point out that as it is a three level townhouse the cleaner had only an hour and half allowed to complete the cleaning and consequently, whilst the service was appreciated, because of the time restriction, it could not be regarded as completely satisfactory.

There was also two hours of respite provided by another person who came every second week and that service was to allow me to have some time away from my duties which normally was used for shopping and other essentials.

I was grateful for these small efforts of assistance and knowing how these things tend to work, I was very thankful because it meant that our foot was in the door and therefore it was very likely that any future adjustments to the assessment would be followed by a more prompt availability of the next level of required services.

In our case we decided to remain in our home in Blackbutt, which admittedly was a tri-level townhouse which made the caring process considerably more difficult than if it had been a single level residence. At this time it was appropriate to discuss in some detail exactly what was involved in making our home suitable and above all safe for Jenny.

I needed to be advised what changes would be necessary to ensure that all parts of the house and even some of the exterior should be made safe for Jenny for as long as her remaining under my care at home was a practical possibility.

It is of course necessary to point out that because of the tri-level nature of the residence the cost involved in the necessary changes was naturally higher however, the type of changes made would be necessary regardless of the size and type of residence in question.

Among some early symptoms the one which caused us the greatest concern was Jenny's uncertain balance when walking and her tendency to fall.

It was a blessing that considering the number of falls Jenny had she did not cause herself any serious injury nonetheless something obviously needed to be done to make the place safer.

At this time it was necessary to consult some people who were in a similar position or were experienced in doing the sort of work to provide facilities to address this particular problem is a process which indicated to me just how inexperienced I was when it came to the needs of disabled people.

Obviously, the changes that we needed to make to our townhouse did not all take place at once but were, as the needs dictated, undertaken during the period of time from the diagnosis of Jenny's condition until it was no longer possible to continue with the appropriate care at home.

Accordingly, whilst I will discuss all of the needs to modify the house and some of its surroundings in this period in the book, they will certainly not all be attended to within the framework of the period.

It needs to be understood that, unless some or all of these things had already been attended to, they would be necessary for the sake of providing your patient with optimum care.

It is recognised that in many cases if the carer is not part of the family or a relation they may not be involved to any extent in these activities but as it is possible for most carers who will be reading this book, they will find it useful advice and understanding.

We were advised to install handrails throughout those parts of the house where, at that stage Jenny was still able to walk unaided thus making sure that she was at all times able to support herself and thereby considerably lessening the incidence of falling.

Jenny was still quite able to make her way from one part of the townhouse to the other with the exception of course of the stairs which, in our house were taken care of by the purchase of stair lifts.

Initially handrails were fixed to the walls from the bedroom, along all walls of the landing opening to the bathroom and the toilet where handrails were also appropriately installed.

It is essential for all these works to be subject to expert advice because quite clearly, it is not necessarily sufficient just to install handrails if they are not exactly at the appropriate height, and in the appropriate places.

This is particularly important in the toilet and the bathroom which will be commented on in greater detail.

The lounge room, the dining area in the kitchen must also feature handrails but, to use common sense, the installation of handrails on every wall would be unnecessary bearing in mind that at that time Jenny was still mentally alert and was able to be told that she should avoid walking anywhere in the house where a handrail was not available.

These were the areas which were attended to in the early stages however, the time would unfortunately come when your patient would begin to have difficulty in attending to personal needs without aid. Greater focus on works that were needed in the bathroom and the toilet made more extensive changes necessary and unfortunately these are works of considerable expense.

Under the appropriate circumstances pensioners may be entitled to a subsidy to the cost of bathroom modifications and as these arrangements appear to differ from one state to the other it is suggested that contact is made with Centrelink who are in possession of all the facts relating to these subsidies.

Some recently constructed homes show an awareness of at least some needs of disabled people in which case some savings could well be made.

Bathrooms, mainly because they are designed to be pleasing to the eye, basic practical considerations may not have played a part in their design.

The area most critical is invariably the shower recess which in many if not most cases will need to be completely remodelled.

In the majority of cases our shower recesses tended to feature glass screens, fixed shower heads and draining in the centre of the floor area.

I must confess I was quite shocked when the plumber presented me with a plan detailing all the changes needing to be made to the shower recess.

To start with the area being enclosed by glass albeit partially, clearly a disabled person who was subject to falling would be presented with quite some danger even with reinforced glass being used. The glass was to be replaced by a shower curtain.

Fixed shower heads have proven totally unacceptable for disabled showers and need to be replaced with shower heads which can be lifted off their mounting and by way of a hose of adequate length may be used by the patient or the carer on all parts of the body.

The shower head is replaced in its mounted position while soap or moisturisers are applied and then once again lifted off to facilitate rinsing of the body.

It is almost invariably recommended that a shower chair be used principally because the patient may not be able to remain standing alone, as well as for the assistance of the carer who then does not need to support the patient and can use both hands in the showering process.

A similar suggestion would also be apt for the area just outside the shower recess because, if the patient can sit down after having been partially dried off in the recess, in order to facilitate the drying of the lower legs and feet, as well as if there is any moisturiser to be applied this can be done by the carer without needing to support the patient in any way.

The next and by no means the simplest or cheapest modification relates to the floor of the shower recess because the central drainage whilst serving the glass enclosed areas quite satisfactorily, will certainly not be suitable to adequately drain the area surrounded by a curtain.

Depending on the condition of the shower floor it will most likely have to be lifted, redesigned to allow the water to drain from the back through the middle to the front of the shower recess, where a drain is installed across the complete front area thus facilitating complete drainage of the area without allowing any water to leave the shower recess and flood part of the bathroom.

As we all know wet tiled floor in any area is a danger and steps may need to be taken to ensure that all areas in both the bathroom and toilet if not already safe, are either re-tiled with a non slip tile or, if appropriate painted with a solution which will provide a non slip surface.

Next, the toilet, which in our house was a separate area, must in addition to handrails already fitted cater for additional aids, and as and when the patient commences to experience difficulties standing up from the toilet, businesses promoting and selling aids for the disabled have a frame which fits over the toilet and will facilitate ease of getting up.

In time there will probably be a need to purchase an extender which fits on the top of the toilet and will become the seat with its own lid, also providing help in standing up and sitting down, assisted by the elevated seat.

A further addition to the toilet, although at considerable expense, would be a bidet, most especially to be installed as soon as incontinence becomes a problem, as an appropriately installed bidet would greatly assist the carer when some considerable clean up becomes necessary.

As part of this modification process it is absolutely imperative that the house is inspected from top to bottom with a view to locating any uneven areas in the floors especially going from one room to the other, locating poorly lit areas where the patient is likely to venture and ensure that the lighting is improved. Loose carpet squares and mats are not recommended and should be removed because they are considered tripping and sliding hazards.

If there are young people in the household they should be constantly warned not to leave loose items around anywhere on the floor of the house such as toys etc., as they also constitute a hazard and could cause the patient to trip and fall.

This peril can be very real even in those areas where the patient is firmly gripping the handrail as a trip can normally cause a spasmodic reaction which more often than not would be the letting go of the handrail in order to extend one's hand and arm to break the fall.

It is essential to understand that these costs to be incurred are necessary and what sometimes makes them even more difficult to accept is the fact that we have no time period to which to relate the cost.

In other words having completely remodelled the house for the sake of the patient's comfort and safety there is no guarantee as to how long the patient will remain under home care and when ACAT will finally classify the patient as requiring "residential high care".

That decision unfortunately signifies that it is no longer practical or for that matter safe, for the patient to remain at home and efforts to find accommodation in a nursing home needs to be considered. There is no point in examining the possibility of providing residential high care at home as the cost on both the physical health of carers as well as the cost of providing additional people to come in because the care in question is 24/7 which, considering penalty rates would be absolutely prohibitive.

Quite clearly no matter how much order to the home environment has been achieved and no matter how much the patient's safety in matters of toileting, moving etcetera is, what ACAT will also need to assess are the number of facilities which exist in a nursing home which do not and in most cases cannot exist in a private house and therefore the level of care which is warranted cannot be given and the change in classification will be warranted.

The cost of being admitted and cared for in a nursing home can be high and more often than not unless the patient is a pensioner, can mean the need to sell the house in an attempt to finance the move.

"WILL I GET BETTER?"

Although it was quite obvious that Jenny was adversely affected by both the growing symptoms of MSA and naturally enough, the gradual realisation of what her future held, quite clearly, life in our household was far from normal. It does not take a great deal of empathy to imagine what effect the sudden loss of Jenny's future meant to us.

To think that those elaborate plans we made just a short while before, our intention to use the initial part of our retirement to tour around Australia, in the main visiting places where I have lived and worked since my arrival in 1949 suddenly became an impossibility.

It is true to say that, as a good carer, I should make an all-out effort to take that period between diagnosis and the end to try and turn it into an acceptable quality of life, sounds quite wonderful but could I take that on any level of sincerity when I knew full well that my Jenny was very difficult to deceive.

There is unfortunately a danger where some successful measures which have been introduced to improve the patient's quality of life, albeit temporarily, could often be interpreted by the patient as a possibility of an improvement in their condition.

Never during her illness did Jenny ever complain nor had anyone ever hear her asking "why me?" however, the prospect of an improvement generated by her better quality of life needed to be carefully controlled. During this period she often asked "am I getting better?" to which all of the people involved in her care would answer saying; "only God knows that".

Jenny did also commence talking of dying and frequently expressed the view of her condition as "this is no life", or "I'd be better off dead". At this point in time it was fairly rare but I anticipated a likely increase over time.

These are the most difficult times when your patient is also the love of your life and no matter how controlled you remain, you experience fear of an event that you have known will inevitably occur.

This is just one example of the ongoing process of grief which would never cease only to be replaced with real grief when the time ran out

Of course we will never know how effective saying that "only God knows" was, when what we fear will finally happen.

That is why I was very grateful for what Jenny managed to get out of the concentrated sessions at the hospital, her time spent with the young safety therapist who, after talking to me found out that Jenny and I were keen tenpin bowlers, introduced Jenny to Wii, an electronic collection of games, including a tenpin bowling game which certainly did capture Jenny's imagination.

I bought a Wii and had it installed on our TV and sound system and a young woman who came to spend some respite time with Jenny two or three times a week, managed to achieve some considerable change to Jenny's attitude. As Jenny was always a very competitive person in sport quite a competition developed between the two, much to Jenny's delight.

At the same time the thought occurred to me that one of the things Jenny used to love to do was to go for a walk in the morning.

Most of the time she would also take Maxine's dog for a walk because Maxine, due to recent knee surgery wasn't able to.

After the accident with the water feature walks stopped and the thought occurred to me that, whilst Jenny's unsteadiness would make the resumption of daily walks somewhat difficult, Jenny was quite steady whenever she used the walker and so, we discussed the possibility of resuming walking limited distances to start with and then see what developed.

To my absolute pleasure I found that Jenny walked quite capably with the walker and of course I walked alongside keeping an eye on her as she did have the slightest tendency to wander from one side to the other.

Because the terrain around our townhouse was not at all suitable for the use of the walker we would drive each morning to Lake Illawarra a short distance away, where a walking and cycle track was the ideal place to do our walking.

We had to make a slight change to the times, because our usual time around eight in the morning found us right in the middle of the path being used by people going to work.

So we changed our time to 9.30 and whilst we still struck some walkers and cyclists, this was yet another experiment which yielded some success.

As we got into this new routine of going to Lake Illawarra I was very pleased to note that as I helped Jenny out of the car and organised her with her walker she seemed very relaxed and as we got going we started carrying on a normal conversation. It was my opinion that removing Jenny from the home environment for any out of the ordinary activity had the tendency of taking her mind off the many things she must have been thinking about whilst at home.

Over two or three of these walks Jenny related to me her holiday in Europe arranged by Contiki with Jeanette some years ago and the fact that the two sisters didn't seem very compatible during that holiday. Quite often their fellow travellers commented that they were sitting at the opposite side of the bus because they weren't talking.

That subject and also quite a number of others discussed by us during these walks were all completely unrelated to the present and therefore served the purpose of replacing in Jenny's mind many of the things that unfortunately resided there most of the time.

In addition to the walks, these and some other ideas relating to some of Jenny's favourite activities such as gardening were introduced and I suddenly realised that keeping Jenny reasonably busy doing things she used to enjoy, and if they were repeated often enough, the theory that, quite regardless of the dismal prospects of the progression of her ailment it was possible to introduce and influence her state of mind giving her at least some interest and therefore some quality of life.

I found that it was terribly important for me as the carer to learn everything about what in any way influenced Jenny's demeanour and the more I learned the more convinced I became that these sorts of actions should be very much a part therapy introduced and practised in the very early part of the illness. The earlier they commenced, it would seem very likely that the result would be an extension of some acceptable standard of quality of life as the patient waits to die.

Most of what is contained in the earliest periods recorded by me, represent in the main a great deal of reading and research as well as experiments involving Jenny in an effort to find possible therapies which are likely to be beneficial if only in the short term. A good deal of the time I spent almost groping in the dark whereas at the time when these lines are read, those reading it will in all probability be already familiar with much of the contents but even if they get a small proportion of new advice or ideas the effort will have been worthwhile.

"Commencing Home Care"

*J*enny having been further assessed, this time as high care, meant that we were the next step away from residential high care which eventually would necessitate Jenny being admitted into a nursing home.

Towards the end of Jenny's rehabilitation program at the local Hospital some of the therapists spoke to me about an Advance Care Plan for Jenny which should be prepared and put in place so that when the time came a lot of the contents of the plan had already been recorded and, although nothing was hard-wired, most of Jenny's instructions would be unlikely to change.

Jenny still at home under my care.

Jenny's condition worsened very gradually and so the increase in level of care although obvious, did not yet reach anywhere near a critical level.

She was no longer able to wander about the house unaided and so we gradually retired walkers except downstairs where, from time to time she would be helped outside to sit at the outside table in the alfresco area or sometimes with a great deal of care she was moved along the edge of the retaining wall where she could reach some of the flowers and give a creditable imitation of gardening.

We also retained the walker in the car as it would be frequently needed to assist if when I needed to take Jenny to the Dr or at times take her shopping to provide her with yet another distraction.

On some occasions we were accompanied by Maxine whose company Jenny always enjoyed and even though I am not a happy shopper, I did enjoy accompanying them listening to the chatter and noting that Jenny enjoyed the experience every time it happened.

Her speech had deteriorated further and I was particularly grateful to the speech therapist who came to see Jenny about every two weeks and each time tested her eating and her swallowing and supervised the consistency of the food which had to be cut very small and it was recommended that she should have more soups.

The stage was also reached where she needed some assistance with feeding although she did manage utensils very slowly and with some difficulty. It was really a matter of allowing her to feed herself for as long as she was happy and then take over and finish as much of the meal as she was able.

This was where the gradual failure of coordination was becoming obvious as, what to us is the normal practice of pushing items of food forward with the aid of one implement onto another held by the other hand was definitely failing.

As I would sit there and observe I knew she needed help but it was one of the many examples where empathy was needed especially being well aware that I was dealing with a person who was very proud and fiercely independent so whatever way in which I offered that help could either please or offend her.

Most of these things in my case were the actions of a novice not quite knowing what the results of any action I decided on were likely to be.

To watch Jenny's struggle with some of the most basic actions could not have been more upsetting.

Watching and attempting to assist a person becoming quite helpless when only recently she had been very keen to assist you is an experience not to be envied.

When I talk about that learning curve, as time went by, I had to realise that the curve would be getting steeper and the learning more and more difficult.

These are parts of the caring activity which really indicate that regardless of which of you has the disease, both of you suffer the consequences.

It was around about this time when Jenny's sister Jeanette offered to come and assist with some care which certainly was welcome.

Another area which had commenced to cause concern was incontinence which started to raise its ugly head in a very small way to start with but like everything else gradually became more obvious.

We were given the name of a physiotherapist in North Wollongong who specialised in incontinence problems and so I took Jenny to see her.

Getting Jenny in and out of the car was also an increasingly difficult task as was her achieving some degree of comfort sitting in the car while travelling.

The session with the physiotherapist turned out to be less than productive simply because, as we were told by the physiotherapist, Jenny should have been sent to her some considerable time before, because the only thing she could do was to provide Jenny with some exercises to strengthen the pelvic floor.

Regrettably those instructions were not really understood and once I got Jenny home the only thing being achieved was that I strengthened my pelvic floor showing and explaining to Jenny how to do the exercise which unfortunately was not successful.

Of course we would have to admit that as the time to attempt these things was decided when the incontinence was so far advanced that it would have been nothing short of a miracle to be able to train a disabled person to learn and practice such an intricate skill.

Once again co-ordination played a role as it does in just about every physical movement and even if Jenny had been capable of reacting to the instructions, parts of her body would probably not.

The necessary changes continued keeping pace with Jenny's condition and so it was decided we must proceed with the changes to the bathroom and the toilet, as well as that we had to obtain supplies of special underwear to make the control of incontinence just a little easier.

By then we had moved Jenny upstairs to one of the spare bedrooms together with the big television from downstairs which then became her daytime room opening out onto the landing from which the toilet, bathroom and the main bedroom opened.

Naturally, no matter how fast the tradesmen worked, this project did throw us out for a number of days making it particularly difficult to continue giving Jenny adequate care, protecting her from the dust, which was not extreme thank goodness for the care exercised by the workers but was present nonetheless.

The commencement of high care did mean increased traffic of people in addition to the ones already rostered.

From the day when the work in the bathroom was completed I gave Jenny her first shower in the remodelled shower recess and certainly found it to be much easier from both the patient and the carer's point of view.

As Jenny's condition deteriorated her showering by a professional carer was rapidly approaching but for the time being I wanted to continue because she was used to me giving her showers after overcoming the usual difficulties that are characteristic to people who have never been showered by someone not even the partner and so the natural modesty made it very difficult on the first few occasions.

Jenny's sister who up until then was an occasional casual helper had commenced to commit herself to looking after Jenny every Tuesday and part of every Thursday in order to give me some respite. I would have the odd quick game of golf as well as taking care of shopping and other duties.

I managed time as best I could and tried to read up any new comments on Facebook. Some in particular surprised me but also made me grateful as they referred to extreme pain being experienced by some MSA sufferers.

I was very surprised and pleased that never to that time had Jenny experienced any pain and upon reflection I can say that that was the case right through the entire period of Jenny's illness. We did however need to be very vigilant that Jenny didn't develop any bedsores considering that she was either reclining, sitting or lying down on her chair or bed which meant she was in danger of developing bedsores practically all the time. These are some of the examples which make some level of training for carers who operate in the home so they are aware of areas of risk not commonly realised.

It was also clear from these comments on Facebook that MSA appeared to be a complaint full of different symptoms and allied complaints none of which were evident in Jenny's case.

That conclusion did in no way interfere with my long and firmly held view that, regardless of each specific terminal or serious illness, they have in the main very similar areas of need so far as providing care is concerned. Therefore I have no hesitation in suggesting that most if not everything contained in this book is by no means exclusive to those who suffer with MSA.

As the new arrangement settled in we had to accept and handle the changes as and when they occurred and our efforts to keep Jenny amused with various ideas became more and more difficult. She was no longer capable of handling the Wii and therefore the tenpin bowling ceased, she was no longer able to read at all however she was still able to look at, admire and quite enjoy coloured pages in magazines.

Attempts to maintain a conversation became quite difficult because Jenny herself became very conscious of the fact that she was unable to carry on a normal conversation which, especially among strangers, became a particular source of embarrassment and it seemed that she would prefer not to speak at all except in short words here and there.

This at times, when empathy was inadvertently forgotten, could cause some annoyance which was most unfortunate because once we realised her tendency to use short words had a very valid reason and not at all what we imagined it to be, we accepted its rare occurrence.

A typical example of this was when every now and again Jenny would say "water", simply meaning that she wanted a drink of water.

I must confess I did on some occasions take that to be a command and I am ashamed to admit that I felt some resentment. It became clear quite quickly that using the word water served the purpose because she got the drink she wanted and it also helped her to avoid trying to construct a sentence which was not only difficult for her but it also became a source of embarrassment.

The state of mind which commenced to exist fairly early in the piece, in fact during the time when we were still able to take Jenny around to visit friends and relatives, was her unwillingness to go anywhere, to meet anyone because of the feeling that she did not want people, most especially those who knew her well, to see her the way she was.

I'm not sure why it is difficult for people to understand this problem but it is and not just among strangers but even relatives with the result that I was often accused by some of Jenny's family of deliberately keeping them away from her.

If only they knew and understood how much more difficult that made my task as Jenny's carer perhaps they would have attempted to develop at least a little sensitivity.

Right from the beginning I intended to remain Jenny's primary carer not because I saw that as my duty as her husband but because of my love for her and the fact that I just did not want to be away from her at any time.

This determination carried on after she was admitted to the nursing home and I needed to get special permission from the management of the facility because as Jenny's primary carer I did have a say over some of the decisions and some of the actions taken by the staff.

I am digressing simply because I think these were comments appropriate in the context of what I'm writing.

While Jenny was still at home I insisted on preparing all her meals which had to be carried upstairs by me and placed on a small table which fitted in front of the reclining chair she was occupying.

She still enjoyed her meals and from time to time we still managed to carry on a conversation providing I remembered the virtue of patience as well as the constant awareness of the need for empathy.

Times commenced to become more and more difficult most especially due to the deterioration of Jenny's condition so far as her incontinence was concerned and as things developed I needed to be very vigilant at night because of the need to frequently take her to the toilet, more often than not clean her up and sometimes even shower her, putting on a clean gown before I put her back to bed.

This of course prevented me from enjoying a reasonable night's sleep as each of such occurrences and quite often two of these occasions represented enough interruption to my rest to have been lost for that night altogether.

I welcomed what respite I was able to have including the two days by Jeanette and then JennyII, *(to prevent confusion)*, the other sister in law, Barry's wife who, committed herself for most Sundays which gave me yet another day's respite.

What probably wasn't appreciated because they weren't aware of it and I didn't speak much of it was that the interruption to my nights was for each of seven nights so, in spite of the fact that I was given three days of respite they didn't include the nights because my duties as a primary carer went on every night.

By this time the helpers coming to the house had taken over the showering which helped me quite a lot, also the washing and ironing was taken out of my hands which did make my days a lot easier.

Of course when we talk about such things we talk only about physical activities and not their inevitable effect on the carers emotions and demeanour. Talking from personal experience in my case as not only the carer but also the husband of the patient, who by the very nature of the caring activity was with the patient needing to observe her day by day as she faded out of this life.

I have referred to this time and time again as the process of having to grieve while my wife was alive and then continue to grieve after she was gone.

I reached the stage where I actually welcomed something happening which required me to jump and attend to some problem Jenny was having which would help to take my mind off other and more depressing things.

The recognition of what are regarded as the many day-to-day actions involved in caring whilst performing the basic caring duties, are by no means the only occurrences which can and do have a sometimes lasting effect on the general demeanour of a carer. This recognition helps to prepare the carer for the emotional impacts that flow from the caring process. I am mindful that by far the majority of the contents of this book relate to situations where the patient and the principal carer are a married couple and consequently some of the emotional turmoil which inevitably is encountered may not be entirely relevant to some of the readers but I am very much of the view that some of the things contained will strike home in each case where the reader's situation very much resembles ours.

Realising that these emotional influences can and often are to the detriment of the relationship between patient and carer, and therefore should as much as possible be recognised and analysed and perhaps more importantly anticipated and discussed. This can go a long way towards negating most ill effects.

I was and still am very grateful for the strength of the relationship between Jenny and I because there had been a number of occasions when we had been tested and it was only because, as progressively difficult as that may have been, our constant communication which continued right to the end, sustained our relationship.

I cannot help but feel that matters such as these rarely form a part of any training which is provided for prospective carers, the absence of which can result in some carers who have undergone training having to face the rigours of caring partially unprepared.

I have spent much time and effort debating the wisdom of my decision that Jenny should be fully informed as to her condition, discussions which took place in Jenny's presence with our David always included all relevant details.

Despite this openness it was some time before Jenny gained full knowledge and understanding of her condition as well as what was in store for her.

All the time, at every opportunity we talked openly and frankly and more often than not we each shed a few tears because of the highly emotional content of our discussion.

One of my fondest and saddest memories, which occurred during one of these talks, when I found myself breaking down and crying and after a while I felt Jenny's hand on my head gently patting it and saying "don't cry".

There is of course no doubt that the overwhelming desire was and would continue to be to remain at home and provide whatever care was necessary in that environment because it would be highly unlikely that Jenny would be prepared to willingly accept a move to a nursing home.

It was at this time when following some discussions with the staff at the local hospital where Jenny had been undergoing some rehabilitation, it was recommended that we should contact Dr Roger Cole who was the head of palliative care for the Illawarra region.

I was advised that I should meet with Dr Roger Cole so that he would have the opportunity of discussing the plan with Jenny and help her to make some of what were likely to be difficult decisions.

As it was always my insistence to be present during such encounters I was very grateful for that opportunity even though it certainly was one of the most painful and disturbing experiences. Nonetheless, I sat and listened to Dr Cole and I was pleased to also listen to comments made by Jenny which suggested to me that this meeting was an appropriate communication considering the subject matter involved.

There was no doubt that Dr Cole was discussing the likely events that may occur towards the end of Jenny's struggle with MSA and once she accepted the inevitable she was assured that she would have optimum care to leave this world in comfort.

I have thought about this conversation time and time again and whilst my mind was at rest because I knew that everything that was said Jenny understood and she obviously came to terms with the outcome, which did help me to accept the prospect which in fact proved to me that up until then although I knew exactly what was in store for Jenny I hadn't really accepted it until I listened to Dr Roger Cole.

As you read on, it will eventually become clear that the carer and the husband or partner sitting through such a meeting may find it relatively easy to accept the subject matter discussed, the most pointed being the patient's decision relating to alternative methods of feeding and of course whether or not the patient would desire to be revived in the case of resuscitation being required.

This was undoubtedly a point at which I needed to use exceptional self-control, as these ideas having been introduced into my thinking, I could feel a sense of fear overtake me and a growing desire to find a way to keep it from happening.

I am not ashamed to admit that crying myself to sleep was not an unusual event.

Here I would comment that those working in health care know that because of the difficulty of decisions involved in these plans they are more often than not put off and possibly never taken care of which, if neglected can and often cause considerable discomfort and even grief.

After much consideration I did decide to include these matters fully in the book, especially because I am hoping that the material contained in this book will be read by inexperienced prospective carers who may never experience some of the emotional turmoil which such things bring about but knowing about them and understanding them I do believe will be important. After all, it is more often than not the carer who brings these matters to their patient's attention thus ensuring as far as possible they begin to be addressed.

Jenny seemed to experience no difficulty in communicating with Dr Cole and she did understand all of what he explained and discussed with her.

Soon after Dr Cole's visit, ACAT called and re-assessed Jenny, deciding to make her subject to "residential high care.

This was, although expected, difficult to take, particularly for Jenny who was not at all ready to accept the verdict that she will have to leave our home.

"*It is Reality not a Choice*"

*T*he time passed slowly from when the final decision by ACAT was given, classifying Jenny as "residential high care", and when the resulting changes were able to be implemented. It took considerable time, because, as was usually the case, there were no vacancies at any of the nursing homes which I visited in my quest to find one suitable.

That task in itself was quite formidable because there being a high number of nursing homes in the area and the surrounds, finding one that in my view was acceptable and also not too far distant from our home was certainly not an easy task.

I was extremely fortunate to have found Ridgeview Aged Care, which was a facility recently taken over by the present owners from its previous operators who had been less than prudent and had been refused registration.

The new owners spent considerable funds on upgrading the facility and when I was first shown over it I found it to be second only to another recently completed nursing home where I had already received advice that as there was a formidable waiting list of patients there would be no vacancies in the foreseeable future.

I found the two women managing Ridgeview to be very nice and most cooperative and whilst on my first visit there were no vacancies it was indicated to me that one or two rooms should become vacant in a relatively short time.

That I suppose, could have been regarded as a blessing because it gave us more time to overcome the strong objections from Jenny which we knew would be forthcoming.

Unfortunately I subsequently came to the conclusion that the resultant stress on Jenny really only served to extend her misery when one considers that the end result was, after all beyond, question.

Her admission to a nursing home was inevitable and so it would have been better if the waiting period was much shorter however, that was not an available choice because in such things we are constantly at the mercy of the prevailing conditions, as well as the generosity of the Government.

Things unfortunately continued to get worse for when we finally received the call that there was a bed for Jenny at Ridgeview and visited the facility to be shown where Jenny was intended to be placed I had to very carefully and diplomatically decline. I explained to the management, that I had a brief chat with the lady Jenny was to share the room with, an elderly lady who was totally deaf and had the television on full volume and was not about to give any consideration to someone else coming into the room, a situation I could not put Jenny into.

I considered myself very fortunate that in view of the long waiting lists existing for these various places the management at Ridgeview was understanding about the situation and said that they would advise me as soon as another vacancy became available.

This actually occurred within about a week and although it was yet another proposition involving sharing a room, it was indicated to me that there was a chance that a single room could become available soon after Jenny was admitted and if that occurred she would be given preference.

With all this to-ing and fro-ing Jenny was becoming quite depressed but thankfully this time when the call came and I went and looked at the room I couldn't find anything that would be a serious enough problem for Jenny to refuse to make the move. I spent some time explaining to Jenny that all of us would be significantly better off if she did accept the offer of the accommodation and it was finally agreed to make the move one Saturday morning.

I now understand that this sort of situation is not uncommon and these are times when one needs to be fairly firm in recognising that no resistance could avoid what was regarded as inevitable.

The preparation for Jenny's transfer to Ridgeview was a highly traumatic event. Thanks to the assistance from Jeanette and Maxine both of whom have been a great help in preparing the things we thought should accompany Jenny although, our townhouse was only fifteen minutes drive from Ridgeview and so this activity wasn't as critical as we thought it to be because we would have been in the position at any time to pick up other items that we may have missed.

Jenny had resigned herself to the situation and consequently was very quiet during all the proceedings and finally on Saturday morning, the 9th March 2013 we took Jenny to Ridgeview where, much to our pleasant surprise there was quite a welcoming committee in the reception area with both the facility manager Michelle, her assistant Cassie and many of the staff including the Chef and her assistant all welcomed Jenny as the new resident.

Just exactly who was more pleased and impressed by this elaborate welcome, Jenny's entourage or perhaps Jenny herself for certainly she did brighten up considerably on arrival and had a brief chat with Michelle who was a person with great people skills and she and Jenny got on very well thereafter.

Jenny was taken to her new room and joined her room-mate who was a wonderful lady in her early 90s, who was a well-known dance teacher in the area during her younger days.

The room was quite full because both patients had a number of family members with them and so Jenny and all of us met the old lady and her two visitors and vice versa and this made for a fairly relaxed atmosphere which, for Jenny's sake I was very grateful.

I had somewhat of a dilemma not knowing exactly how soon a single room would be available and therefore we decided just to get a few pictures and perhaps one item of furniture to create a touch of home for Jenny and resolved to work on that over time as appropriate.

Our arrival at Ridgeview was early to mid afternoon and therefore Jenny's first meal was going to be the evening meal which, I was told was usually served just after 5 PM.

Every resident had the option of going to the dining room, (depending on physical condition) or the meal would be served in their room.

By the time Jenny was admitted, details such as preferred diets, allergies and other important information essential for her safety had been discussed and recorded however, I knew that in Jenny's case this would be a constant ongoing process depending on what we observed to be necessary as well as instructions from time to time from Katie, the speech therapist who did continue her regular visits.

At this stage Jenny was still capable of feeding herself slowly with the bulk of the time taken up by her picking up the appropriate utensil(s), she was still needing some assistance from time to time.

Where Jenny needed some assistance at that time was the periodic manipulation of food on the plate where I observed that she was using the spoon or fork to manipulate the food towards herself which then needed to be regularly moved back towards the centre of the plate to prevent food from spilling.

After dinner Jeanette departed and Maxine and I stayed behind for a while to make sure Jenny was settled.

The two room-mates actually struck up a small amount of conversation which suggested that Jenny would have no difficulty in relating to her new friend.

It is strange that those of us who have had no business or dealings with nursing homes or even hospitals, tend to expect a very quiet and sedate atmosphere which is very quickly dispelled after all, when you have a lot of staff some of whom are young women, interacting during the carrying out of their duties, there was bound to be some discussion, instructions and even some light hearted banter breaking any silence there may have been.

Maxine and I left at about 8 o'clock and as Maxine came in my car we went home together and found very little to talk about.

I sensed that Maxine was more than a little depressed about things even though she was probably one of the most active of all the people around Jenny and was very strong on advocating the ultimate benefit of Jenny's move to a nursing home. I knew that Maxine used me in her arguments pointing out that I was struggling to keep up with the pressure and that this move would at least give me some relief.

After saying goodnight to Maxine I put the car away and went inside the house to be greeted by Chloe our little long-haired tabby who was to become my sole home companion for the time being at least but quite possibly for as long as we both shall live.

I find it very difficult to describe arriving home to an empty house and I was wondering if I was indulging in a dress rehearsal of what was eventually to come.

I had decided right at the outset of starting this book that, as difficult as it was proving to be I would relate simply everything, including any emotional turmoil like the one I was certainly feeling then.

I wasn't feeling like having anything to eat but soon after I came inside there was a knock on the door and it was Maxine with a plate of cold cuts and salad saying that she had a feeling that I would find use for it.

I thank God for seeming to have found support at the right time and certainly from the right person, no wonder I called her our guardian angel.

It is amazing that even if you are regarding yourself as religious, when times are difficult the urge to pray tends to intensify and I recall spending most of that night in prayer.

I found myself having to face reality on my own and that was difficult beyond the extreme simply because I had not yet reached the full realisation of just exactly what was happening to us.

Jenny, who was one of those people who never ceased to be concerned about the problems of others and continually went out of her way to offer assistance where she felt it was needed, to be struck down by one of the cruellest neurological conditions certainly did not appear to be fair.

I was grateful for the opportunity to be allowed to continue in the role of primary carer for Jenny not because I was in any way expressing concern about the staff at Ridgeview being able to fulfil their responsibilities but because I wanted to be with her as much as was physically possible for the entire duration of her illness.

We had taken Jenny's wheelchair with us to Ridgeview which I think was appreciated as there were likely to be periods of shortage of those implements.

The first two days went quite smoothly with Jenny eating relatively well and not really showing a great deal of discomfort about being where she was but also because of her newly struck up friendship with her neighbour she was fairly busy talking to her as well as the daughter who was visiting for most of the day.

Day three was fairly critical to Jenny's well-being because during the night the dear old dancing teacher was called to teach dancing to angels and notwithstanding the discretion being used in removing her little friend from the room Jenny was quite devastated.

I needed time to pacify Jenny and I found myself saying something I always hated to hear others say; "She is much better off where she is", based of course on an assumption that we know exactly where she is, although in this case that assumption probably was reasonable.

Two days later when the granddaughter and her husband came to remove some of the lady's belongings the granddaughter spoke to me at some length and said how much they admired Jenny even for the short time they had known her.

She said that both prior evenings Jenny kept enquiring about how she was and if there was anything she could do which amazed the two visitors.

The granddaughter said; "Here she was being in such a terrible position which would be enough to depress anyone and yet she was taking time to be concerned about an old lady she hardly knew".

There was nothing extraordinary about that because it was typical of Jenny as a person.

The granddaughter of the old lady came and visited Jenny regularly during her stay at Ridgeview.

It was only a matter of two days before Jenny's new room-mate appeared.

She was a Russian lady who was unfortunately unable to speak English notwithstanding the fact that she had been a resident in Australia for some years.

She had a mystery illness which was subsequently diagnosed as a stroke, was unable to use her left limbs and Jenny was filled with pity for her but at the same time was frustrated because of the inability to communicate.

She was an example of absolute and deplorable neglect on the part of her family. We never saw her husband because he never called, she had two grown-up sons and we only very rarely saw one of them for a very brief and rather impersonal visit. A matter of a few weeks later, Cassie came to advise that the room right next door to the nurses' station was being vacated and that they would be moving Jenny in there within one or two days.

I was very pleased to receive this news but typically Jenny was more concerned about what was to happen about her room-mate rather than being pleased for herself.

While the room was prepared I organised a nice display cabinet with some of Jenny's favourite knick-knacks, as well as a small table and a small refrigerator which was put on the table.

I bought a DVD player because we were both very fond of music and I knew that she would especially enjoy listening to Andre Rieu and his orchestra. Maxine's son Kurt mounted a small television on the wall for Jenny to be able to view from her bed.

Over the ensuing period I must have purchased every available Andre Rieu DVD and CD and they did prove to be a wonderful diversion for Jenny whenever she needed a lift.

We were never short of good music because in addition to the ones I purchased both Maxine and Jenny II were in the habit of buying some discs for Jenny.

After the first few days I was regularly putting Jenny in her wheelchair and wheeling her around the facility which had some lovely gardens both within the complex and also in the outer grounds. In doing this I found that the standard wheelchair was not satisfactory for Jenny because she had difficulty keeping her head up and as there was no headrest at the right level so her head kept leaning back awkwardly.

Jenny in the garden at Ridgeview showing the
accumulating effect of the condition.

I spoke to Michelle Megson and she put me onto the man
representing a business in Wollongong which specialised in items for
the disabled and he recommended a wheelchair made in Sweden which
he felt would be perfect for Jenny.

There wasn't one of these in stock and so it was brought in from
Sweden and it was brought to Ridgeview for Jenny to try.

These are the times when on the one hand as a carer of a disabled
person you're very keen and anxious to find things that are suitable and
will make your charge a little more comfortable. On the other hand you
have to face up to what these things can cost and I found this particular
item priced well over $2000.

The thought occurred to me at the time that the need for these
things doesn't change from one disabled person to the other, but the
ability to provide them with what they need with the cost sometimes
proving quite prohibitive, can be difficult.

I was grateful that the chair did suit Jenny and we were able to
purchase the item which served Jenny right up until the date of her
death.

Ever since the cruise I remained convinced that activities as well as various items of interest were very important in displacing the negative frame of mind which undoubtedly plagued Jenny from time to time. This new wheelchair from Sweden was very manoeuvrable and being much larger than normal wheelchairs was extremely comfortable.

It had large pockets on both sides of the unit which became very useful for Jenny's copious numbers of tissues and also the MP3 unit which I obtained for use in the sessions of self hypnosis as well as the earphones fitted into the pocket quite comfortably so when they were not in use they were always handy in the pocket of the chair.

Being in a room of her own was I believe very beneficial for Jenny. She was no longer needing to consider another person in the room and so she could have the volume up a little louder, she could have a snooze if she felt without having to listen to the sound of the television of her room-mate in short, she was able to enjoy solitude.

When there wasn't much doing by way of things to fill her mind there was constant movement of people and to take advantage of all that excitement I often placed Jenny in her big chair and wheeled her out into a position where she could be left right next to one of the thoroughfares and as people came past many of them stopped and talked to her which pleased her no end and thus served as a satisfactory diversion.

Quite regardless of where Jenny was stationed in her chair it was easy to use the MP3 and the earphones when one of the sessions was due and even though there was a fair amount of passing traffic, Jenny could happily relax, usually with her eyes shut listening and except for some cases, most of the passing traffic walked by without disturbing her but if they didn't, Jenny was usually in a hypnotic trance and therefore ignored them.

I was someone else for whom the new arrangement was very beneficial because I suddenly found that a lot of the pressure was removed after all, much of the work I was doing at home was now being done by the staff at Ridgeview, leaving me free to be with Jenny, or during times when Jenny slept, which was quite often, I was able to carry on with notes for this book and other things like reading and using my little Chromebook yet another present from Paul, to do some research on the Internet having been given permission to hook onto

Ridgeview's Wi-Fi. As well as the book, I also continued to work on the program of self hypnosis, using notes provided to me by one of the many people involved with hypnosis I maintained contact with on the web.

I continued in my determination to find, to experiment with and to apply if the experiment has proven worthwhile, any measure which would in some way extend what remained of life with a view to making it worthwhile to the patient and provide her with a reasonable quality of life for as long as that was possible.

There is no doubt in my mind now that the move to Ridgeview for Jenny was beneficial and even though she was gradually getting beyond the point where any physical activity like the Wii game, going for walks, doing many things she loved doing and so we had to make an all-out effort to ensure that her mind was kept as active as possible and use whatever was at our disposal to achieve that.

Some of this was being achieved purely by Jenny finding herself in a community of like people where she was not only able to relate but also found a lot of people who she perceived to be worse off than she was.

Having made the commitment to continue as the primary carer of Jenny I felt myself becoming a member of the community of Ridgeview, of course not quite to the point of being a resident.

There were occasions when I wasn't feeling like reading or working on material recently researched, and Jenny being asleep or being involved in one of her self hypnosis sessions I often walked through the high care section where Jenny was resident, and met some of the incumbents finding that there is a great deal to be learned from hearing some of their experiences.

The example I mentioned earlier concerning Jenny's Russian roommates' family members, I observed being repeated fairly frequently where it almost seemed as if some people brought an elderly family member to be admitted, never to be seen again.

Some others who treated visiting a relative who was a resident as purely an obligation or, it is sad to say, because of the likelihood of gaining some financial advantage when the relative finally passed on.

After a while I started to explore the possibility of some external excursions and in particular whether or not it was feasible to take Jenny home to visit and of course to see her little Chloe who was certainly missing her.

I found information on the availability of wheelchair taxis and at first I thought I struck an obstacle because I was told that the Swedish chair was too large to fit into these specially modified taxis.

Cassie mentioned that she heard there was a larger taxi which was capable of taking large units comparable in size to a stretcher out of an ambulance and so I traced this larger unit and true enough it had no problem accommodating Jenny's chair.

We also discussed the possibility of using the wheelchair taxi to take Jenny to Mass once a week however, much to my surprise Jenny was not keen to contemplate that.

Considering that our parish priest held mass at Ridgeview once a month for those residents who wished to attend, Jenny was always keen to attend and looked forward to having Holy Communion.

When Jenny was still well we used to attend vigil Mass each Saturday evening at the parish church in Albion Park, and this was a practice I continued even when Jenny was unable or unwilling to join me and, thanks to Father David, I was permitted to bring holy communion to Jenny on each of those days.

I was rather ashamed that I didn't discern the answer for her reluctance to attend mass outside of Ridgeview, after all it was no different from her being reluctant to go anywhere outside Ridgeview where she would be meeting people who knew her and would see her the way she was.

I also found that there was no problem with me taking Jenny in the chair outside the limits of Ridgeview and so we used to go for a walk around the block which took us past the neighbouring school which, just inside the fence, had some poultry maintained by some of the small schoolchildren and Jenny used to get me to stop and she watched the fowls as they fossicked around the grounds.

One of the most memorable excursions using the combination of the Swedish chair and the large wheelchair taxi was on a Wednesday when we managed to coordinate a most enjoyable event involving Anne and Alan, (Alan being my stepson and Anne his wife), visiting from Canberra and by way of purely a coincidence, Ann, who was our gardener attending the garden once every third Wednesday, was also present that day.

Originally of course the suggestion was purely to meet Alan and Anne at our townhouse for a change by taking Jenny there mid-morning and arranging for our visitors to be there at the same time.

Fortunately the gardening was just about all done by the time we had all arrived and because during these visits it was not possible for us to take Jenny and her chair inside the house simply because of the size of the chair, as well as the difficulty caused by the stairs we settled down in our alfresco area which occupied the corner of the garden with the now famous water feature as the backdrop.

That area was our favourite, sometimes just to sit there at any time of the day when conditions were pleasant, or indeed having our meals including our breakfast and several times our dinner.

We had a barbecue which had a wok facility which of course was right up Jenny's alley and consequently I recall having many virtual feasts, sometimes for just two of us and many times for friends who used to enjoy the experience.

We naturally invited Ann to join us and she stayed with us for lunch. It was a pleasure to see Jenny enjoy the company and we were rather amused by the fact that Jenny seemed to have made a special effort to make Alan eat more and in fact she became quite adamant to ensure that he was being looked after.

It was a pity that Maxine was otherwise occupied because normally whenever I took Jenny home Maxine could always manage to be there and depending on the time of day and indeed depending on the day quite often she managed to have her two granddaughters in attendance which pleased Jenny no end as she loved both those girls.

Jenny was always very pleased to see Alan and Anne whenever they visited.

These were clearly the type of episodes which provided Jenny with a welcome change which invariably caused her to brighten up considerably.

Shortly after that event, I received a recommendation from David Barker to meet Dr Michael Barbato who is a Dr involved in palliative care purely in a voluntary capacity as David thought that in addition to having consulted Dr Roger Cole it would be useful for Jenny to communicate with Michael.

The visit was arranged and Michael called to see Jenny at Ridgeview however, I think Jenny may have found that the subject was adequately covered with Roger Cole and didn't seem very receptive to Michael's visit.

I would have to say that whilst I found that embarrassing I was delighted to strike up a relationship with Michael which led to a number of meetings we arranged at Blackbutt.

During these events we talked at length about many things including palliative care especially considering an essay I was working on at the time comparing the various ways available and being discussed and promoted centering around the options of various ways of ending life.

We also talked about Jenny's illness and experiences relating to it and I found these meetings were settling me down at a time when I was not as relaxed or as focussed as I should have been.

Perhaps we tend to forget that such counselling type sessions are sometimes welcomed by the carer and that certainly was true in my case.

These meetings with Michael Babato brings to mind someone else who had been a wonderful help to me in my hour of need namely Wal Edwards a good friend whom I met in 2007 at a golf tournament who at that time was about to celebrate his 91st birthday.

We were both entrants in the International Masters Golf Championship (a name conjuring up a standard of golf which neither of us conformed to), held on the Gold Coast and I had to marvel at this man's demeanour and stamina as at that time (and to this day) he was a grief counsellor as well as very actively involved in the New South Wales Veterans Association.

Jenny and I had visited Wal and his wife Joan at their beautiful chalet in Kangaroo Valley and we became good friends. Wal celebrated his 99th birthday in November 2015 and was mentioned in that year's honours list.

I recall mentioning earlier in this book the belief that when a couple, be they a marriage or a partnership, are unfortunate enough to find one of them diagnosed with a terminal illness, both share all that goes with such a condition but not both in the same way. This applies most especially if the partner not subject to the disease devotes themselves to the role of carer.

Any opportunity that is available for such an individual to communicate with someone who is understanding and possesses the necessary people skills to provide the appropriate counselling must be invaluable and should be welcomed if not pursued.

Yet another friend with whom Jenny had a very strong relationship from earlier times was Heather who up until some 7 or 8 years ago was married to Peter, Jenny's younger brother but that union failed and they divorced sharing the care of three young sons.

Jenny had known Heather since Heather left school and at that time Jenny had a restaurant and employed Heather until such time as she commenced her training as a nurse.

Although there were more than 20 years between them they became very firm friends and even though Heather lives in Canberra she managed to visit Jenny at home and later at Ridgeview fairly regularly and her visits had a highly beneficial effect on Jenny.

Jenny's condition was slowly but obviously deteriorating most especially her coordination with eating, her speech and her moods tended to fluctuate, being quite bright at times but more often exhibiting what appeared to border on depression.

To the credit of Ridgeview there were regular sessions of physical exercises, tai-chi and other group sessions, as well as occasional visits from outside entertaining entities all of which helped a great deal to take troubled minds off, and break the monotony, of life in a nursing home.

Early during her stay Jenny did attend tai-chi classes each week and seemed to enjoy them but it was quite obvious that in time she would likely lose the ability to participate simply because any such activity must depend on coordination.

Two additional services available to residents both of which were welcomed by Jenny were; three manicurists from a local beauty shop came to the nursing home on the first Tuesday of each month and they were available to whoever felt in need of their services and provided a full manicure for $10.

The second service was a hairdresser who was available once a week for the various services provided by hairdressers always at a discounted cost.

Jenny was especially keen on the manicure because having looked after her hands herself, she found herself losing that ability because of the gradual loss of coordination she was most grateful to have her hands looked after by an expert.

Jenny had beautiful hands and her nails always looked marvellous although her sister Jeanette was often somewhat scornful because of the length of the nails and also sometimes Jenny's choice of colour.

Because of these two services being available Jenny always looked nice and it is no secret that good appearance to any individual has an uplifting effect.

Being very anxious to enable me to continue to improve the self hypnosis sessions with Jenny since her admission I commenced to write new scripts concentrating in the main on positive thinking and putting more accent on relaxation in preference to the type of content that was typical of the dream sessions which I had been advised against.

There is really nothing miraculous about someone learning the ins and outs of self hypnosis after all it is a simple process and the patient undergoing self hypnosis was really just simply listening to a recording specially prepared to be appropriate for the particular situation.

Thanks to my son Paul he continued to set up what really amounted to a recording studio which then enabled me to continue to write the script, then read it into a microphone and record the content, after which the facility to vary the level of the tone of the background music to be barely audible in order to prevent the sound from overriding the voice and therefore interfere with concentration.

I was provided with several sample texts, some actually recorded on MP3 format so that I could learn the way in which these scripts were composed and also the recommended volume and intonation which lends itself to a person allowing themselves to be hypnotised.

I was at first concerned that it would be inappropriate to use my voice for Jenny to self hypnotise but I was assured that not only would it be appropriate it would be highly recommended. I was told that because

Jenny was familiar with my voice and trusted it, that it would be the ideal medium to be utilised.

I discussed this intended program with Michelle Megson and she suggested that it seemed impractical for me to spend so much time writing and recording when I should be resting and sleeping. Yet I seemed to have ample time during the day when Jenny was resting or sleeping or perhaps even looking at magazines.

Michelle suggested that as they had some vacant rooms upstairs in the building she would recommend to her principals that I be allowed to use one of the rooms to do the work on the proposed self hypnosis program which would effectively allow me to transfer my equipment into the room at Ridgeview and leave more of my evenings free at home after I left Jenny for the night.

Whilst I was very grateful for this suggestion I pointed out to Michelle that I would need to discuss it with Jenny just to make sure she had no problem with it.

To my delight Jenny not only had no objection but she expressed pleasure at the fact that I could be doing my work which after all was for her benefit in the same building, in fact in a room almost straight above her room and because I proposed to show Jenny how she could use her mobile phone to summon me if there was a need, she could be perfectly safe. Naturally, it wasn't the intention to spend much time over long periods, possibly only a half an hour to an hour at any one time.

This proved to be an excellent move and I was very grateful to Michelle that she could arrange it.

As I have stressed previously, I should continue to make it clear that it is now just as it has been for some time my opinion, that this program as well as some others similar to it should be implemented as soon as possible after diagnosis in order for them to be effective.

In Jenny's case, soon after we commenced these sessions some time before Jenny's move to Ridgeview, some quite obvious benefits commenced by way of an improvement to Jenny's general demeanour but more importantly as these sessions were improved and were applied as frequently as three or four times a day, each of them lasting around 30 minutes, the time during which she was under their influence she was prevented from having depressing thoughts and being pensive and negative.

I don't for a moment believe that hypnotism in any form is likely to provide a cure for any of these serious neurological conditions but as I indicated previously it is my view that as carers it is not our role to be looking for or researching possible cures but rather to be spending our time and resources on any ways and means which we believe will be effective in providing comfort to those under our care and help to extend that period they have remaining of life to contain as much quality as possible.

Since Jenny came into Ridgeview I had a great many opportunities to observe and sometimes talk to other residents and whilst the purpose of this book was initially and still is to relate everything to do with Jenny's battle with MSA, it would be wasteful not to note and refer to things I've learned simply by observing things and happenings.

In my opinion it is essential for carers to read and learn about all things around the practice of caring simply because it will improve their skills and in turn improve the lot of those under their care.

One of my observations of general behaviour of people around disabled persons, and this was in particular prevalent at Ridgeview, was that unfortunate human tendency of speaking to people differently than you would normally speak to others just because they're circumstances are different.

We all know that this is particularly true in the case of people addressing little children they do it by turning into little children themselves.

They also do it when speaking to foreigners, they slow down their speech and raise their voice sounding as if the person they're talking to is hard of hearing.

I do believe that nothing would give more comfort to sick and disabled people than being treated as normal, talking about normal subjects perhaps laughing when it's appropriate not when you feel compelled, in other words normal everyday communication is not only appropriate but essential.

Jenny and I have repeatedly discussed many of these things and notwithstanding the fact that having a discussion with Jenny was becoming increasingly difficult, we loved the interaction just as we always did and that I would recommend to everyone.

This I believe is the area where most of the reluctance on the part of a disabled person to participate in a great deal of social activity occurs, because quite simply there is no pleasure in being treated as someone different and made to feel in need of special consideration.

At this time our most enjoyable period was when I took Jenny out in her wheelchair either to one of the gardens or right out of the limits of the nursing home where Jenny could observe the traffic, sometimes see the children coming out of or going into the school, in short, observing what we knew to be the normal world. I would thoroughly recommend this to be a very important activity for carers especially looking after an individual patient because it is one of the best ways of schooling yourself in the practice of empathy.

Watching Jenny observe the poultry running around in the school-yard, the longer I watched the more I started to see what she was seeing and feel the way she was feeling.

It is amazing how we are able to gradually suppress our own needs and wants when our mind is focused on the needs and wants of another.

In Jenny's case I heard people talking to her while visiting her, and as they departed, telling me or telling each other about how she must be feeling and I find myself wondering, "is this an attempt at empathy? " or is it just an attempt to show sympathy for the sake of just saying something.

Have we ever wondered, how is it possible for a person to completely immerse themselves, to actually slip inside the skin of someone in order to achieve what true empathy really means?

Quite clearly, there are circumstances which can make the true and full practice of empathy quite impossible. How could you possibly put yourself in place where a person is suffering the total atrophy of all systems and how would you express yourself?

A similar question would be appropriate when dealing with complete dementia or an advanced case of Alzheimer's.

It is important to realise these things when dealing with people who are quite helpless and perhaps it is appropriate to remember that simply talking to someone expressing patient and kind thoughts and perhaps accompany that action with holding their hand or perhaps giving them a gentle massage to their hands and/or feet would, I am of the view, be most therapeutic.

Similarly, a simplified form of self hypnosis with appropriate music could if nothing else, induce some level of relaxation and if it does we have achieved some good.

Why do we tend to be reluctant to visit someone who is ill and who would probably appreciate a visit from us because we know them well we have even regarded each other as friends but we hesitate because "what will I say?".

I don't intend to apologise for keeping the subject of empathy to the fore indeed, I intend to continue to do so.

During her most recent visit Katie the speech therapist expressed some concern as to the obvious deterioration of Jenny's speech as well as some indications that her chewing and swallowing of food had also deteriorated.

This was discussed with the nurse on duty and we were advised by Katie that she would like to see all food cut into very small pieces and as well as that, Jenny's food should be administered by a teaspoon making certain that no excessive lots of food were taken at any one time.

These instructions were passed on to the chef and the next meal was as ordered by the therapist.

It was some days later when Jenny experienced her first serious attack of choking and we immediately summoned the nurse on duty and fortunately the substance wrongly swallowed was able to be dislodged but of course we were left with the residue of stress and upset caused by the event.

As it happened Maxine was visiting and I don't believe I have ever seen Maxine turn such a pale colour having obviously been badly upset.

The greatest concern of course was for Jenny who experienced the panic that goes with a choking attack and when you are already in a badly depleted physical condition anything causing such panic could have serious consequences.

Seeing and spending time with Jenny every day tended to make the deterioration of her physical condition much less obvious but whenever we held hands which was very frequently, I felt her hands and fingers being nothing but skin and bone and similarly her arms and legs which suggested that it would be difficult if not impossible for her to lose any more condition.

My discussions with Michelle and the nurses increased, in order to attempt to keep ahead of changes which were likely to occur.

One of these was the obvious need of the use of a lifter which was looming, as Jenny's ability to maintain her posture even sitting in her chair was obviously ebbing.

This of course tended to bring with it the need to make some changes in the manner in which the patient is handled because naturally there would be increasing weakness in joints and even in the bone structure and consequently actions which seemed natural and normal to assist the patient were no longer appropriate.

For example when there is a choking attack if this occurs early in the illness period, a hug from behind using some force often assists in dislodging the substance causing the choking. If that method was used with Jenny's condition at this time it would be highly likely to have caused a fractured rib or two which of course would be highly undesirable.

Similarly the habit by handlers to assist the patient from one posture to the other, for example from the bed to the chair and vice versa, by placing arms under the armpit of the patient which, even though the patient is a lot lighter due to the loss of condition, could result in a dislocation or other injury of the shoulder.

These and many other similar perils would not be part of the repertoire of an inexperienced carer and if such episodes were not part of the initial training or, which is quite possible, they may have been forgotten, problems can and do occur.

One of these is the matter of the arrival of visitors most especially members of family, who as the patient's condition deteriorates tend to increase in numbers and without being unkind, this in many cases is due to a matter of regret for perhaps having neglected their sick relative prior to his or her present condition.

One of my problems used to be that Jenny's relatives with the possible exception of Ray who had to travel a long distance from Queensland, and of course Jeanette who was never regarded as a visitor, was definitely the tendency to turn up without notice which never failed to amaze me and indicated that when it comes to empathy some people just don't have any. These things are very difficult to control because if you chastise people for doing the wrong thing they take offence without giving some reasonable thought that maybe they are just being chastised for a good reason.

There have been many instances where Jenny's bed was surrounded with well-meaning relatives talking amongst themselves while Jenny was either asleep or making out she was.

No matter how ill Jenny was some people who came to visit gave her a great deal of pleasure.

A visit by Eve, my sister in law (the widow of my late brother) was always a highlight although rare because of the distance from Canberra

The main one of these was without question Maxine, whose genuine friendship and love for Jenny was without question.

I have already mentioned Heather who made Jenny's face light up every time she walked in.

My son Paul, his wife Judy and my three grandchildren were also welcomed very warmly as was Alan my stepson and Anne his wife.

Visits by Jenny's brother Peter when accompanied by his three sons, the presence of the boys always pleased Jenny even though these visits were steadfastly unannounced.

Jenny's two nieces Tracy and Chanelle were especially welcome as they were both very much loved by Jenny and I should point out that none of the above except Peter, ever turned up without being announced.

Jenny was now having some fairly frequent but minor choking episodes all of which were successfully prevented from being serious however, it needs to be appreciated that what we may be classed as "minor choking episodes" can easily be minor episodes of actual aspirations of small amounts of food accumulating in the lungs, causing an infection and turning into pneumonia.

I found it necessary to ask the therapist to come and re-examine the situation and the result of that was that Jenny's food was to be pureed.

Naturally she continued to be fed with a teaspoon and the spoon was not to be filled.

This of course meant that meals took a lot longer and because I felt it inappropriate to tie up a staff member Jenny was in the main being fed by Jeanette, when she was there, Jenny II most Sundays and of course dear Maxine whenever she visited during mealtimes and other than that I was doing the feeding.

For the time being Jenny was still eating relatively well but interestingly enough I noted that she was very decisive about what she wanted or didn't want.

If Jenny shook her head on the arrival of the next spoonful there was no earthly good to try and coax her to just take that last mouthful because she wouldn't.

The next choking episode occurred on a Saturday when unfortunately I was upstairs and it was a nurse who heard Jenny coughing and she rushed in to see her and at the same time had me paged and I was able to be there within minutes.

I found myself quite puzzled because it certainly wasn't meal time, I wasn't aware of anything that Jenny had within her reach which she could have eaten but nothing we were able to do stopped the coughing which was pretty constant and fairly severe.

The nurse then insisted on summoning an ambulance and Jenny was taken to Wollongong Hospital and I followed in my car.

A thorough examination wasn't able to detect any foreign matter and so the Dr surmised that the problem may have been caused by an accumulation of saliva however I suggested to him that no matter how accumulated it was it was unlikely that saliva should remain obstructing the windpipe in preference to going down into the lung.

Nonetheless I stayed with Jenny until 1 o'clock on Sunday morning and as the Dr suggested that she may be kept in overnight for observation I went home.

I returned to the hospital early on Sunday and when I couldn't find her in emergency I went up to the ward where she was not to be found. I went back down to emergency and a nurse on interrogating the paperwork advised me that Jenny was returned to Ridgeview at 2:30 AM. To this day I have not been advised as to why neither the hospital or Ridgeview telephoned me to advise what had happened.

As Jenny's condition deteriorated, the nurses and I decided that it was time to do yet another and perhaps a final review of her Advance Care Plan.

Jenny was fully aware of what was being discussed and we went through the care plan principally to make sure that we confirmed with her the two main stipulations, the first of which being her refusal to accept anything other than normal natural sustenance and that in the event of a need to resuscitate it was not to be done.

The decision to use a hydraulic lifter to move Jenny from the bed to the chair from the chair to the bathroom or toilet was confirmed and here is yet another reason for me to be grateful to Paul and Judy who originally bought the stair lifts to be installed in Ridgeview, and also purchased a hydraulic lifter for use for Jenny at Ridgeview.

I was aware that this was in no way done because there was a shortage of the equipment at Ridgeview because that was not the case but I was grateful that equipment such as that could be used for Jenny which was her equipment.

I felt that it was perhaps that last touch of independence which she would have appreciated.

These turns of events made me realise that recognising what appeared to be the inevitable would require me to make the appropriate arrangements which, without any doubt were best to be attended to well ahead of time. Whilst these things would be terribly difficult any time, leaving them to be taken care of after the event would be a great deal more difficult.

Our parish priest David whom we regarded as a personal friend to both of us was under treatment for cancer but before he commenced the chemotherapy he advised me to turn to a young woman who was running things at one of the larger funeral parlours and who impressed David with her understanding and people skills. Hence I sought her advice on what actions to take.

Not surprisingly it was very sound advice.

I was relieved to find that this lady took most of the worries off my shoulders by asking some simple questions relating to the practical aspects of funeral arrangements and I found that not only was I able to make the arrangements for the service and the cremation but also the arrangements for the service to be held at the parlour's Chapel and in the absence of David she even managed to arrange a priest. I am not sure whether people reading this book will find it inappropriate for me to have mentioned all of this as surely, these would not be commonly attended to by a carer but I believe that because I was the carer and the husband and this in part is my story it is appropriate and it is not out of the question that other carers will be in a similar situation.

Shortly after the inexplicable choking episode came the real thing. Notwithstanding the fact that Jenny was being fed a pureed substance there was obviously a fairly substantial accumulation of food in her mouth something the therapist did mention as one of the possibilities, and it was a matter of not just simply swallowing but possibly the need to dispose of the accumulation in preference to spitting it out. An ambulance was called and Jenny was taken to Shellharbour hospital and I accompanied her in the ambulance

The Drs in the hospital said that there was a likelihood of an infection developing fairly quickly and after prescribing antibiotics both the specialist and his assistant came and spoke to me by Jenny's bed.

I'm not sure what Jenny understood because she was very stressed.

The specialist made it clear to me that he was looking to me for a decision as to whether or not they should turn to some aggressive treatment.

I said I was not prepared to make a snap decision and they left saying that they would be in the ward for some time to come yet. The truth be known it was simply that I didn't understand what the specialist meant by saying "resort to aggressive treatment".

A little while later I walked down the ward and saw the specialist's assistant looking at some paperwork at the nurses' station and so I asked to see him. I had a feeling all along that this young man was someone I could talk to and seek advice from.

After he explained that the aggressive treatment would mean the continuation of antibiotics ignoring the hopelessness of Jenny's condition, he then explained to me that in their opinion Jenny had reached a stage where these episodes of aspiration would become more and more severe and that in reality they're at a point of no return. In other words, he said we would be keeping Jenny alive without there being any hope of her condition changing to anything but worse.

He pointed out that atrophy is not an event that has ever been known to reverse nor has it been known to reduce in frequency nor in severity.

He then advised me that Dr Roger Cole was due in the ward later that day and he said that if I felt Jenny would like to see Dr Cole he would arrange to hold back Jenny's discharge to Ridgeview until Dr Cole had a chance to talk to her. I went back to the ward and spoke to Jenny and she said she would like to talk to Dr Cole and so we waited.

I found myself in the position where it was now my turn not to want to speak because I knew that I would break down because I felt that the weight of all this was getting too much to bear.

We held hands as we had almost constantly done of late and I recalled that during the recent few days Jenny talked about wanting to die.

I know now that she said that more often to Maxine than to me and without doubt that was because she knew that Maxine would tell me.

Roger Cole came in and sat on a chair right next to Jenny and explained precisely what the young Dr explained to me and we both had a feeling that Jenny knew exactly what he was saying and she agreed with everything.

She was told that because she had reached a stage in her illness when short of a miracle there was nothing that could be done and so between the nurses at Ridgeview and Roger Cole's own staff, Jenny's care would be taken over but continued at Ridgeview with absolute certainty that she would not suffer and that the end would come without pain and in complete comfort.

It was late afternoon by the time Jenny was returned to her room at Ridgeview and I stayed with her until very late that evening, not wanting to leave her and at the same time being very apprehensive about going home.

Notwithstanding the fact that the final outcome of Jenny's illness was not in doubt ever since that fateful diagnosis, every time something indicated that we were coming closer to the inevitable end it became an almost unbearable shock.

Looking back now I realise that even though I was well aware of what was progressively happening, deep down there was always a small spark of hope that she would recover and live.

Each time we talked about and introduced something that seemed a good idea at time and even though we knew that even if it worked it was not something that we could ever hope to be a cure.

It is very difficult to put into words the feeling I had each time when I looked at Jenny lying there knowing that this terrible disease will continue on its way relentlessly taking Jenny away from me.

Making arrangements for the funeral was not something I anticipated but it was the practical thing to do notwithstanding what emotional turmoil it resulted in.

"PALLIATIVE CARE – THE END IS NIGH"

he previous day's time spent at the hospital and the events which took place I found very difficult to relate to and come to terms with.

Notwithstanding the fact that both Jenny and I had been through all the relevant details on the subject of what was ordained to happen at the end of the fight with the disease, as was the case very often during her illness it was necessary for us both to come to terms again and again with the relevant truths.

My previous night was almost completely without sleep and I spent hour after hour contemplating my conversation with the specialist, our meeting with Roger Cole as well as the few words I spoke with two of the nurses on duty on our return to Ridgeview.

These are the sort of occasions where anyone who is married to the patient and also the primary carer, one is inclined to be facing an experience similar to that of having a split personality.

It is pertinent for me to point out that I am writing this more than a year after Jenny's death having had to abandon a number of efforts because of the emotional toll it took on both personalities.

I have not been able to make up my mind just to what extent my attempt at preparing an essay on euthanasia had influenced my thoughts when considering Jenny's fate however, I do believe that the fact that I was in possession of a fair amount of knowledge and understanding about the various options involved, did assist to a point.

I knew that for religious reasons, I would always be totally opposed to the practices promoted by the proponents of euthanasia in Australia because without doubt their proposal involved the deliberate taking of a life no matter how valid the reason.

Palliative care did unquestionably present itself as the only alternative where those offering it do so stating that it is a way of allowing nature to take its course.

I appreciate that this is a point often under dispute but those disputing it no doubt try to ignore that whatever assistance is provided in this process it is provided to ensure that the final moments are peaceful and without any pain or discomfort.

Whether or not the method used; namely the progressive increase of the dosage of injections of morphine constitutes the taking of a life is probably the opinion by many however it is not unreasonable to accept that in this case life had for all intents and purposes ended when the decision to accept palliative care was made.

To me, what I felt was essential to both understand and accept, was that the successive events which needed to bring this decision to the fore was recognised and advised by expert medical opinion and accepted by the patient and the carer.

Those who will be reading this book and will hopefully never have to be involved in this sort of decision-making, reading about it will certainly not cause them any harm.

On the other hand, if the reader is a primary carer, the involvement in this process will be similar to that which I endured and accordingly I feel it is important knowledge because if this book is to introduce as much reality as possible to the reader, then hopefully the knowledge and the understanding so gained will prove to be beneficial.

I got to Jenny's room early next morning remembering that I was advised by the nurses on the previous evening that the instruction "nil by mouth" was noted on her file and that the kitchen was advised accordingly.

Here was yet again something with which I needed to come to terms.

A little later that morning a nurse from Roger Cole called and spent some time in the nurses' station followed by a visit into Jenny's room where she had a brief chat with Jenny again assuring her that she was under the best care and that all would be well.

She then had a talk with me reiterating what had already been explained that once a patient accepted palliative care, the need for sustenance ceases however she did point out that on rare occasions when that wasn't the case, sustenance could re-continue at the patient's request.

The support to which Jenny had become accustomed in addition to my being her primary carer continued, so Jeanette continued to attend on her usual days and that on any occasion when either of us was absent for one reason or another a staff member would always be within easy reach.

Jenny's room being in near proximity to the nurses' station was a blessing.

The practical benefit to be gained by the readers of this book from this point on narrows somewhat as on the one hand, the usual range of duties and activities expected of a carer reduced substantially and on the other, the effect emotionally on the partner suddenly deepens and so, it's up to the reader to read on or not, remembering however, that what follows is a very human experience and even if it doesn't strike a chord, it certainly cannot harm.

Under these circumstances as much as it may be desirable by either the partner, the primary carer or for that matter anybody else involved to ask "how long to the end?", is, to put it plainly, not possible to answer as not one instance will necessarily resemble another.

At this point in time there was still some ability to communicate but by and large we had been reduced to holding hands and continued to be aware of each other's presence by feeling the odd squeeze from time to time.

Once again as a number of times before, I was very reluctant to speak except on rare and perceived to be necessary occasions because of a fear that my emotions would take control.

I should have realised and possibly did, that I should not have applied that sort of restraint because Jenny of all people was acutely aware of my many weaknesses and me being excessively emotional certainly was one.

It was important that some music preferably one of the many numbers that Jenny was obviously fond of should be playing softly in the background and for the time being a self hypnotic session was still used however after a day or two only music remained. At this time and since the last choking episode Jenny was on oxygen and she was to remain so for the remainder of her time.

There were still a number of things I had to attend to, there was still our home to worry about, bills to be paid but all of these things I did when Jeanette was present to ensure that Jenny would not be unattended.

I felt for some time that Jenny was able to speak if she wanted to but for the first two days since returning to Ridgeview she made no such attempt.

Then, very late one afternoon she said a few words which were difficult to understand.

What I understood I found strange and very moving because she appeared to be saying; "I don't want to leave you, you will be defenceless".

The first part of the sentence didn't surprise me but I wasn't sure why she thought I would be defenceless until long after she was no longer there to explain.

Each evening before I left Ridgeview, I would say some prayers while holding her hand and I did feel that she prayed with me in silence.

When we finished praying I could feel her squeezing my hand.

Every now and again I came out of Jenny's room and just walked along the corridor of "High Care" and I found that a lot of the staff would stop and talk to me all of them expressing sorrow and concern over Jenny's condition.

I remember one of the senior nurses' aides telling me that her colleagues frequently talked about Jenny and how her selfless attitude and her constant concern for the welfare of others had left a lasting impression on all of them.

She told me that she herself and many of the others had expressed a desire to start developing a different attitude to their job.

When I returned to Jenny's room I tried to convey these sentiments to her and there was a smile on Jenny's face and a squeeze of my hand.

It is so very difficult to cast my mind back over that period and virtually re-live what I went through as I sat by her bedside, holding her hand and listening to her breathing in short and fairly snappy breaths, listening to the hiss of the oxygen and all that, backed by some of her favourite music.

I noted the monotonous repetition of the necessary action under palliative care, the nurses coming in and checking on the syringe driver and make whatever adjustment was deemed to be necessary.

This was, as I remember, the part of the principles of palliative care which caused me some concern.

I was grateful for the fact that I was consulted each time an adjustment to the dosage was made but as I remembered, this was the procedure which ensured that there was no pain or discomfort and to the best of my knowledge, Jenny gave no indication of any pain or discomfort.

Even though I was not aware of any pain Jenny experienced for the entire period of her illness up to her decision to accept palliative care, what I needed to think about and accept was that nobody was really aware of any likelihood of pain felt by any patient undergoing palliative treatment. The question remains that notwithstanding the complete absence of pain to that time, is it not likely that the cessation of any sustenance being administered would, in the event of the complete absence of the treatment with morphine, surely have been likely to occasion pain?

Spending all that time by her side, I had plenty of time to think and my thoughts as they relate to this procedure involving the syringe driver made me decide to record my thoughts, as I am certainly of the opinion that they may be of interest and perhaps even of some value to the reader.

If I am correct in assuming the point at which the control of the care of the patient passes to palliative care represents the decision that the life of the patient should cease then, the period over which that process should take should not be allowed to be excessive.

My reasoning for this is that it seems to me that palliative care as it is practised today takes account of all considerations necessary to ensure the safety and the comfort of the patient however, little if any consideration appears to be given to the welfare of the carer/partner or indeed to the well-being of persons, near and distant relatives who may be desirous of attending the bedside of the patient for the duration of his or her progress to the end.

Notwithstanding the fact that Jenny undoubtedly had made up her mind that she did not want to live and that passing away was to her an escape from a life which she didn't consider to be worth living, once the palliative process commenced, Jenny seemed determined to hang on and continue to do so for an extraordinary period.

From the point of view of both being her carer and also her husband if anything could be done to shorten that waiting period, such a decision would not have any effect on Jenny simply because the constant administration of morphine in increasing quantities would have prevented any such perception to exist.

After quite a number of days I felt it was time for me to arrange for a priest to come and administer last rites.

Father David being ill and the priest relieving at the parish not being available I found myself making some telephone calls in an effort to find a priest who would come to Ridgeview to attend to Jenny.

The first and logical place would have been to call the Bishop's house and there I was given the parish priests' telephone number and finally I was able to speak to someone at the parish office in the parish of Kiama where the parish priest was on leave but the assistant, Father George was prepared to come and advised that it would be late that afternoon because of other commitments.

Father George arrived at dusk accompanied by another priest both of them working in Australia from India.

Before going into Jenny's room we sat down in the visitors lounge next door to the reception area where Father George asked to be told something about Jenny as he said it was very important for him to at least get to know something about the person he was about to administer last rites to.

He pointed out that because Father David was on sick leave and was ill, he would naturally have been the ideal person to attend because of his relationship to Jenny and I but because of his illness he was unable to attend..

I appreciated all that and agreed. After all there is something very personal as well as spiritual about the ceremony that was about to be undertaken by Father George assisted by his friend.

On entering the room I tried to arouse Jenny as much as I thought that to be possible, but I was not successful and whilst obviously alive which we could tell by her breathing assisted by the oxygen she gave no sign of life and therefore Father George decided to continue without any obvious participation from Jenny. I recall Father George saying to me as we were leaving Jenny's room that he felt that Jenny was well aware of our presence and that he was convinced that, as much as was possible, she was part of the ceremony.

I was very grateful to Father George and as I'm writing this, I am living on the Central Coast of New South Wales, over 12 months since Jenny's passing, I received a number of telephone calls from Father George one on the anniversary of Jenny's death, and, as I recall soon after my move to the Central Coast.

I thank God for people like Father George.

We had regular visits from the nurse from Dr Cole, making sure that all was well and I found that Jenny was by no means the only patient at Ridgeview under palliative care.

I noticed Jeanette talking to the nurses frequently and I knew that she was being kept up to date on exactly what was happening and I was very grateful for that because although I tried very hard not to show it I was very close to breaking down at every opportunity.

For me to be able to talk to Jeanette or for that matter members of the family,which I felt I was expected to do, would have resulted in a very tearful session on my part and somehow I felt prevented from doing that.

Similarly I was sure that as some of Jenny's family came and went during this period they would have been updated by Jeanette and I think she would have recognised that as part of her role as both Jenny's sister and one of her carers.

There were a couple of times during the night when I was at home I received a call from Ridgeview from one of the nurses who advised me that as there had been a significant change in Jenny's breathing patterns it was appropriate for me to be present.

Nothing significant happened on both of these occasions but we all felt that Jenny was reaching the end of the road and so, we waited and it was that waiting which was without doubt the worst period of my life.

I held Jenny's hand almost incessantly except when the nurses came to check on her but even holding her hand contained no reward.

I was thinking about having passed my 84th birthday a matter of two or three weeks before, and the fact that I was 15 years Jenny's senior, why wasn't I lying there in the bed and Jenny holding my hand?

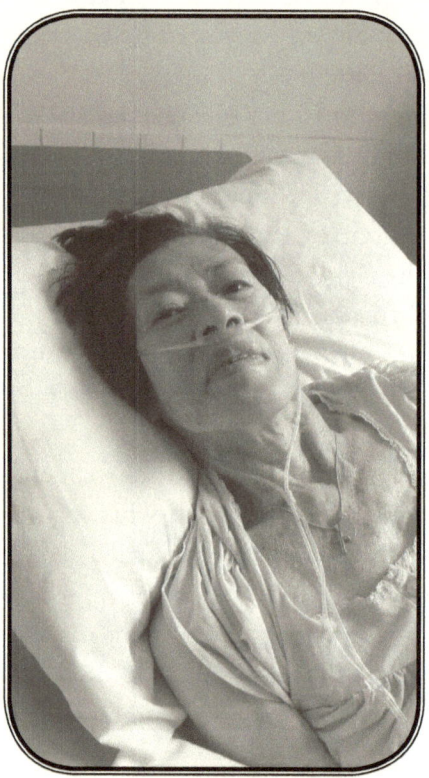

Sadly<there were indications that the inevitable was nearing>

Maxine arrived and took my place by Jenny's bed and I knew that she would want to hold her hand and did so.

Jeanette who sat alongside me stayed where she was and I moved to the other side of the bed and held Jenny's left hand.I recall Jeanette talking to me but I cannot remember what she was saying.

It was then when we detected a significant change in Jenny's breathing and in a little while Maxine said "she's gone" and as I was holding her left hand, I felt her spirit leaving her body.

Somebody called the nurse I think it may have been Jeanette, and the nurse verified that Jenny had passed on.

I noted gratefully that at Jeanette's invitation I was left alone with Jenny. How can I explain what it felt like to go on holding her hand in the desperate hope that her spirit may still be there.

I have no accurate recollection as to just how long I sat there but I do remember begging her not to go even though knowing that her spirit had gone and the thing I feared had happened.

The nurse who was attending to Jenny came into the room and said that Dr Barker was on his way.

I reluctantly left the room and I noticed that Jeanette went in and I felt it was appropriate for the two sisters to be together for a while.

David Barker breezed in and went into the nurse's station to attend to whatever paperwork needed to be completed. I found myself surrounded by members of the staff expressing their sympathy, many of them actually crying and rightly or wrongly I couldn't help but feel that at a time like that sympathy may well be appropriate and appreciated but for me, being left alone would have been preferable.

CONCLUSION

*I*f we can strip down to the barest facts this account of Jenny's illness you have just finished reading, I hope in my heart that it can be regarded as serving the purpose of the book.

I am not regretting or apologising for having given a full account of all that happened including my emotions and feelings which to me, were very much an integral part all that I felt both as a husband and as a carer.

I cannot expect that every reader will react to the contents in the same way simply because most if not all of the circumstances would be different and if I can suggest, the possibility that the depth of feeling that existed between Jenny and I may or may not apply, I feel sure the contents will be understood.

The period of time immediately after Jenny's death is still and will continue to remain a blur and thankfully I had my son Paul and his family giving me strong support, in time resulting in my move from Blackbutt to Lake Munmorah on the Central Coast where I am close to Paul and his family who of course I now regard as my family.

I sincerely hope that the contents of this book will serve a purpose to some if not all its readers especially those who are contemplating caring for whatever reason, to look after someone they love and who is very ill, to accept the call to caring as a step along the way to the many disciplines within health to treat it as a calling and a career.

Figures recently published indicated a staggering number of people regarded as carers but just under what terms of reference I didn't note.

In New South Wales alone going by figures taken some four years ago put the number of carers for the State at in excess of 800,000 of which some 8% were children caring for elderly family members. The fact is, that just as the ageing population is substantially increasing so unquestionably, will the need for people providing care also increase.

I appreciate that a lot of the subject matter covered and discussed in this book which were not at all or barely known and understood during the period covered by the book, by the time you read it, most of those early examples would have become an integral part of the practice of caring but conversely there are and will continue to be many new questions relating to therapies and treatments which are hitherto not known and understood but hopefully an effort to research them, test them and apply them will continue in the interest of easing the burden of serious diseases and the fact of ageing itself.

March 2016.